# DIHYDROGEN BONDS

# DIHYDROGEN BONDS

## Principles, Experiments, and Applications

**VLADIMIR I. BAKHMUTOV**
Texas A&M University
College Station, Texas

**A JOHN WILEY & SONS, INC., PUBLICATION**

*Library of Congress Cataloging-in-Publication Data:*

Bakhmutov, Vladimir I.
  Dihydrogen bonds : principles, experiments, and applications / Vladimir I. Bakhmutov.
      p. cm.
  Includes index.
  ISBN 978-0-470-18096-9  (cloth)
1.  Dihydrogen bonding. I.   Title.
  QD461.B225 2008
  541'.226—dc22

                                                                    2007024689

Printed in the United States of America

10 9 8 7 6 5 4 3 2 1

*To my wife*

# CONTENTS

# PREFACE

Among the various attractive forces holding molecules together, hydrogen bonds are the most effective, due to their pronounced directionality and relatively low bonding energies, which are particularly important for noncovalent supramolecular synthesis and crystal engineering. It is clear that intermolecular hydrogen bonding has a profound impact on the structure, stability, and stereochemistry of inorganic, organic, organometallic, and bioorganic molecules and molecular assemblies built via hydrogen bonds. Despite the relative weakness of hydrogen bonds (commonly estimated as 5 to 7 kcal/mol) due to cooperativity, they are responsible for the spontaneous formation of the three-dimensional shape of proteins and for the double helix of DNA and other complex molecular aggregates. In some sense, intermolecular hydrogen bonds act as glue in the buildup and design of molecular crystals. The main advantage of hydrogen-bonded crystals is the fact that they are weak and energetically flexible enough to allow annealing and editing. On the other hand, they are strong enough to impart stability to crystal systems. The role of hydrogen bonding is also well recognized in proton transfer reactions, where hydrogen bonds act as organizing interactions.

Hydrogen bonding, one of the oldest fundamental concepts in chemistry, is constantly evolving, due to the appearance of new experimental and theoretical methods, including new approaches through computer chemistry. Dihydrogen bonding is the most intriguing discovery in this field. Although ideas about the interaction between two hydrogen atoms with opposite partial charges have been exploited by chemists for a long time, formulation of this interaction as a bonding between two hydrogen atoms was first suggested in 1993, at which time dihydrogen bonds become objects of numerous theoretical and experimental studies.

I have been involved in these studies since their beginning, collaborating closely with Robert Crabtree and Jack Norton in the United States, Robert Morris in Canada, Rinaldo Poli and Odile Eisenstein in France, Maurizio Peruzzini and Claudio Bianchini in Italy, Agusti Lledos in Spain, Lina Epstein in Russia, and other scientists who have provided great contributions to the development of this interesting field. In this book I summarize these results through the concept of dihydrogen bonding.

Although this book is a scientific monograph, not a textbook, some parts of the book could be incorporated in general courses for senior undergraduate students, graduate students, and postdocs. The material in the book is based on numerous original works and recently published scientific reviews and therefore reflects the current situation in this field as objectively as possible.

The classification, energy, geometry, and dynamics of hydrogen bonds as a part of weak noncovalent interactions are described briefly in the first two chapters. Hydrogen bonding is considered in terms of modern criteria based on the topological aspects of electron density. In Chapter 3 a general view is formulated on the concept of dihydrogen bonds, focusing on the nature of bonding and on its energy and geometrical features. Experimental approaches to dihydrogen bonding in the gas phase, in solution, and in the solid state are discussed in Chapter 4, where the readers' attention is concentrated on x-ray and neutron diffraction and on IR and NMR experiments. Some experimental methods for the determination of bonding energies for dihydrogen bonds are also a focus of this chapter. In Chapter 5 we look at intramolecular dihydrogen bonds C–H···H–C, C–H···H–B, N–H···H–B, and O–H···H–B in various classes of chemical compounds, from very weak bonds to quite strong ones, in metal hydrides. The role of dihydrogen bonds in the stabilization of molecular conformational states and in dehydrogenation reactions ends the chapter. In Chapters 6 and 7, based on numerous experimental and theoretical data, intermolecular dihydrogen bonds formed by hydrides of various chemical elements and proton donors in solution and in the solid state are discussed. Chapter 8 is concerned with correlation relationships between energetic, structural, and electron density parameters established for various dihydrogen bonds. Proton affinity and basicity factors are applied to characterize quantitatively proton-accepting strengths of hydridic hydrogens with respect to regular organic bases. Some aspects of dihydrogen bonds in their application to supramolecular chemistry and crystal engineering are covered in Chapter 9. In Chapter 10 we consider mechanistic aspects of proton transfer to hydridic hydrogens in solution and in the solid state and discuss kinetic methods and experimental and theoretical approaches to investigations of various dihydrogen-bonded intermediates and transition states lying on reaction coordinates. Some energy aspects of proton transfer are also considered.

Each chapter concludes with a section that highlights details concerning intra- and intermolecular dihydrogen bonding, the nature of the elements donating hydridic hydrogens, the role of dihydrogen bonds in molecular aggregations, stabilization of molecular conformations, and finally, in proton transfer reactions, where dihydrogen-bonded complexes can appear as intermediates or transition states. In the final chapter we sum up, focusing on two general conclusions: how short or long a dihydrogen bond can be, and what environmental factors act particularly strongly on dihydrogen bonding.

Thus, the book covers the nature of dihydrogen bonding, factors controlling its energy, its role in molecular aggregations, and movement of a proton along a dihydrogen bond, ending in full proton transfer. The book should therefore be of interest to scientists working in the areas of materials, supramolecular structures, self-assembly, hydrogen storage, acid catalysis, and homogeneous hydrogenation catalyzed by transition metal complexes.

Finally, I would like to thank Dr. Ekaterina Bakhmutova-Albert for useful discussions and technical assistance.

VLADIMIR I. BAKHMUTOV

*Department of Chemistry*
*Texas A&M University*
*College Station, Texas*
*May 2007*

# 1

# INTRODUCTION: WEAK NONCOVALENT INTERACTIONS

The development of chemistry in the last 20 years has revealed a significant shift of interest on the part of theoreticians and experimentalists [1,2]. Earlier, chemists' attention was concentrated on atoms and atom–atom bonds. This strategy has been very successful in the creation of new molecules with unusual structures and with new chemical and physical properties. However, two decades ago, the primary objects of chemical studies become intermolecular interactions leading to complex molecular assemblies that exhibit unusual and often unique macro properties. This situation has dominated in all areas of modern chemical science: from physical, organic, inorganic, and organometallic chemistry to material science and biochemistry, and has resulted in the formulation of new chemical disciplines: supramolecular chemistry and crystal engineering.

As is well known, molecular assemblies can be created due to secondary and weak interactions referred to as *noncovalent bonding*. This term implies that such bonding *does not lead to the formation of new two-electron σ-bonds*. In this context, the formation of new σ-bonds symbolizes strong interactions that change molecular skeletons and therefore require significant energy. In contrast, *noncovalent synthesis* proceeds through the formation and rupture of secondary interactions between molecular subunits. Since the covalent skeletons of initial molecules are not affected, this synthesis occurs at the other end of the energy scale [1]. On this scale, small energies often act against entropies that complicate the synthesis. However, if noncovalent interactions are cooperative, thermodynamically spontaneous supramolecular aggregation finally provides a significant energy gain.

If the utilization of weak noncovalent interactions leading to molecular aggregations is a general principle in supramolecular chemistry, and periodicity is a general prerequisite in the crystalline state, then periodically distributed noncovalent interactions constitute the basis of molecular crystal engineering [1]. In other words, molecular crystal engineering can be considered as supramolecular solid-state chemistry, again based on weak noncovalent interactions.

*Dihydrogen Bonds: Principles, Experiments, and Applications*, By Vladimir I. Bakhmutov
Copyright © 2008 John Wiley & Sons, Inc.

Finally, mastering secondary noncovalent interactions is important not only for supramolecular chemistry and crystal engineering, but controlling these forces is fundamental in the context of an understanding of complex biological processes, particularly the principles and mechanisms of molecular recognition [1–3].

The attractive forces that can hold molecules together include van der Waals interaction, electrostatic attraction (when molecules are charged or polar), and hydrogen bonding. Since there is no clear border between a very weak hydrogen bond and van der Waals interaction, the latter requires some explanation.

It is well known that neutral and nonpolar molecules push each other away when the distance between them is small but are attracted to each other at longer distances. This idea was suggested in the second half of the nineteenth century to rationalize the kinetic behavior of gases. Then van der Waals formulated an equation according to which molecules in the gas phase undergo the influence of an attractive force field. The latter explains the existence of the condensed phase as a result of intermolecular attractive forces. These interactions are very weak and can be estimated as $10^{-2}$ to $10^{-1}$ kcal/mol per one van der Waals contact. At the same time, these interactions are long, act at long intermolecular distances, and form a van der Waals potential energy surface. Intermolecular potentials cannot be measured directly, but they associate with the spectra of van der Waals complexes: The intermolecular modes of the van der Waals complexes depend directly on the potentials that hold these complexes [4,5]. It is worth mentioning that when other intermolecular attractive forces are weak, the van der Waals modes, being very soft, have large amplitudes. Their frequencies can be estimated as a *few tens* of reciprocal centimeters for complexes containing nonpolar monomers [5]. By comparison, frequencies in hydrogen-bonded systems are a *few hundreds*. Experimentally, the van der Waals modes can be detected directly by laser-based far-infrared (IR) spectroscopy or as sidebands in mid-IR and ultraviolet (UV) spectra. IR spectroscopy of cold gases can also provide useful information on van der Waals complexes.

Molecular systems formed by van der Waals forces can usefully be characterized by zero-electron kinetic-energy photoelectron (ZEKE) and resonance-enhanced multiphoton ionization (REMPI) spectroscopy. For details, readers are referred to ref. 4, where the complexes formed by phenol and argon are of particular interest. In fact, these spectroscopic methods can show transitions from van der Waals interaction to hydrogen bonding. Figure 1.1 illustrates a van der Waals complex, phenol • Ar, where the distance between the argon atom and the center of the aromatic ring is as long as 3.58 Å. The structure shown in Figure 1.1(b) corresponds to a hydrogen-bonded complex. Ab initio calculations have predicted the existence of several isomeric structures for this phenol • Ar system, two of which are characterized by the lowest energies (Figure 1.1). In contrast, REMPI and ZEKE spectra only support van der Waals structure existing under supersonic jet conditions.

Among intermolecular attractive forces, hydrogen bonding is the shortest in terms of intermolecular distance but the most energetically strong. In fact, a hydrogen bond can provide an energy gain per structural unit from 2.4 kcal/mol to

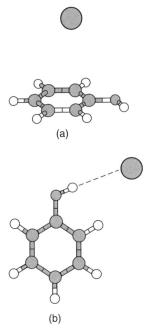

(a)

(b)

**Figure 1.1**  (a) Van der Waals complex formed by phenol and argon molecules; (b) corresponding hydrogen-bonded complex. (Reproduced with permission from ref. 4.)

24 kcal/mol, or even more in the case of charge-assisted hydrogen bonds [6]. By comparison, the dipole–dipole interaction energy, for example, between even the very polar molecules HCl is only 0.8 kcal/mol. It is obvious that hydrogen bonds as well as dihydrogen bonds are most perspective for creation of supramolecules from molecular subunits.

Hydrogen bonding is one of the oldest and most fundamental concepts in chemistry [7]. Hydrogen bonds are found in solids, in liquids, and in the gas phase and often dictate aggregates states of chemical compounds, classical examples being $H_2S$ (gas) and $H_2O$ (liquid) molecules. The hydrogen bonds define the crystal packing of organic and organometallic molecules [2], modulate the reactivity of groups within molecules, and stabilize conformational and configurational molecular states [7]. The role of hydrogen bonding is well recognized in the stabilization of complex biological macromolecules, enhancing the selectivity binding of substrates to these biological molecules. Finally, hydrogen bonding plays a very important role in proton transfer reactions.

The concept of hydrogen bonding is constantly evolving from classical hydrogen bonds to nonclassical (or nonconventional) hydrogen-bonded complexes. Here, on the basis of new experimental and theoretical data and new approaches to this problem, the nature of a proton-donor component and a proton-acceptor site is reformulated completely. In addition, experimental criteria that have been used successfully earlier for the detection of hydrogen bonds are also changed.

For example, due to their elongations, the *red shifts* of vibrational modes observed for proton-donating bonds (e.g., O–H, N–H) were the strongest evidence of a hydrogen-bond interaction. However, since the discovery of *blue-shifting interactions* [8], where the proton-donating bond is *shortened* upon complexation, the experimental formulation of hydrogen bonds has changed completely.

Among the various nonconventional hydrogen bonds, dihydrogen bonds are the most unusual and intriguing. These bonds are formed between two hydrogen atoms, the first positively charged and acting as a proton-donor component, the second negatively charged and acting as a proton-acceptor site:

$$H^{\delta+} \cdots ^{\delta-} H \tag{1.1}$$

The first reliable data on the existence of such bonds within crystal structures came from the mid-1990s [9], when Wessel and co-workers reported on the unusually-short intermolecular contacts B–H$\cdots$H–N. Nowadays, studies in this fast-developing field involve a large number of chemical elements participating in the bonding quartet X–H$\cdots$H–Y. It is clear now that this unusual bonding plays an important role as an organizing interaction in molecular architecture (intramolecular dihydrogen bonds), in molecular aggregations (intermolecular dihydrogen bonds), and in the reaction ability of molecules.

The aim of the book is to show (1) a diversity of dihydrogen bonds; (2) the nature, geometry, energetics, and dynamics of these bonds; and (3) the factors that control the energy of dihydrogen bonds from very weak to very strong.

Since it is very important recognize the place of dihydrogen bonding among various noncovalent interactions and to show differences between hydrogen and dihydrogen bonds as well as their similarity, we begin Chapter 2 with a brief description of hydrogen bonds, their classification, their energy, geometry parameters, and dynamics. For details the reader is referred to numerous scientific reviews and monographs, among which refs. 10 to 19 are notable.

## REFERENCES

1. D. Braga, F. Crepioni, *Acc. Chem. Res.* (2000), **33**, 601.

2. G. R. Desiraju, *J. Chem. Soc. Dalton Trans.* (2000), 3745.

3. J. G. Planas, F. Teixidor, C. Vinas, M. E. Light, M. B. Hursthouse, *Chem. Eur. J.* (2007), **13**, 2493.

4. C. E. H. Dessent, K. Muller-Dethlefs, *Chem. Rev.* (2000), **100**, 3999.

5. P. E. S. Wormer, A. van der Avoid, *Chem Rev.* (2000), **100**, 4109.

6. F. Hibbert, J. Emsley, *Adv. Phys. Org. Chem.* (1990), **26**, 255.

7. G. A. Jeffrey, *An Introduction to Hydrogen Bonding*, Oxford University Press, Oxford (1997).

8. P. Hobza, Z. Havlas, *Chem. Rev.* (2000), **100**, 4253.

9. J. Wessel, J. C. Lee, E. Peris, G. P. A. Yap, J. B. Fortin, J. S. Ricci, G. Sini, A. Albinati, T. F. Koetzle, O. Eisenstein, A. L. Rheingold, R. H. Crabtree, *Angew. Chem. Int. Ed.* (1995), **34**, 2507.

10. H. J. Neusser, K. Siglow, *Chem. Rev.* (2000), **100**, 3931.

11. L. A. Curtiss, M. Blander, *Chem. Rev.* (1988), **88**, 827.

12. A. C. Legon, D. J. Millen, *Chem. Rev.* (1986), **86**, 635.

13. M. Meot-Ner, *Chem. Rev.* (2005), **105**, 213.

14. P. A. Kollman, L. C. Alle, *Chem. Rev.* (1972), **72**, 283.

15. I. Alkorta, I. Rozas, J. Elguero, *Chem. Soc. Rev.* (1998), **27**, 163.

16. M. D. Joesten, L. J. Schaad, *Hydrogen Bonding*, Marcel Dekker, New York (1974).

17. S. Scheiner, *Hydrogen Bonding: A Theoretical Perspective*, Oxford University Press, Oxford (1997).

18. S. J. Grabowski, *Annu. Rep. Prog. Chem. Sect. C* (2006), **102**, 131.

19. A. Kohen, J. P. Klinman, *Acc. Chem. Res.* (1998), **31**, 397.

# 2

# BRIEF SUMMARY OF HYDROGEN-BONDED SYSTEMS: DEFINITIONS AND GENERAL VIEW

Despite the seeming simplicity of the task, in reality it is not easy to say what a hydrogen bond is. The various definitions suggested in the literature show that a unique formulation of the hydrogen bond does not exist. For example, the extended Pauling's definition, "a hydrogen bond is an interaction that directs the association of a covalently bound hydrogen atom with one or more other atoms, groups of atoms, or molecules into an aggregate structure that is sufficiently stable to make it convenient for chemists to consider it as an independent chemical species" [1], seems to be the most appropriate for chemists working in the field of crystal engineering. In fact, two factors, association and stability, are the focus of this definition. However, it is well known that hydrogen bonds are dynamic in nature, and hydrogen-bonded complexes are often unstable, populating an area on the potential energy surface that does not correspond to an energy minimum. Desiraju provides another definition: "The hydrogen bond is a three-centre four-electron interaction that is stabilizing and directional with certain spectroscopic attributes and reproducibility of occurrence" [2]. However, as we will see below, this definition is suitable for strong hydrogen bonds that have a covalent character. Moreover, these spectroscopic attributes are not always available, and sometimes they do not provide an accurate formulation of the interaction (e.g., red- and blue-shifting interactions). Therefore, to avoid general definitions of hydrogen and dihydrogen bonds, we prefer to show them in tables and figures following rather phenomenological definitions.

## 2.1. CONVENTIONAL HYDROGEN BONDS: THEORETICAL AND EXPERIMENTAL CRITERIA OF HYDROGEN BOND FORMATION

The world of hydrogen bonds, always very diverse, became more so after the discovery of dihydrogen bonds and dihydrogen interactions. It should be pointed

*Dihydrogen Bonds: Principles, Experiments, and Applications*, By Vladimir I. Bakhmutov
Copyright © 2008 John Wiley & Sons, Inc.

out that these two types of bonding are not the same, as we will see below. Despite this variety, manifested in their geometry and bond energy and also in their dynamics, particularly in solutions, they can be classified generally as conventional (or classical) or as nonconventional (or nonclassical), where the difference is based on the nature of a proton-accepting site.

A hydrogen bond can be called *conventional* or *classical* if it is formed between a partly positively charged hydrogen atom in proton-donor component $X-^{\delta+}H$ (where X is an electronegative element such as O, N, or halogen) and the lone electronic pair of electronegative element :B acting as a proton-accepting component:

$$X -^{\delta+} H+ :B \rightleftharpoons X-H \cdots :B \qquad (2.1)$$

X–H is often termed a *hydrogen bond donor* and B, a *hydrogen bond acceptor*. Since the formation of a hydrogen bond requires an orientational preference, the interaction in eq. (2.1) is not simply an electrostatic attraction between positive and negative charges, despite the fact that this interaction provides the main contribution to the total energy of hydrogen bonding (see below). On the one hand, this feature is in contrast to other noncovalent interactions, and on the other hand, it approaches classical hydrogen bonding to covalent chemical bonds.

An interaction can be taken as a *conventional hydrogen bond* if its formation corresponds to the following criteria [3]:

1. The interaction energy is measured from weak to medium.
2. interaction is accompanied by a remarkable interpenetration of isolated electronic clouds of the two moieties involved.
3. interaction results in a certain electron transfer between the two moieties.
4. interaction exhibits a preferred geometry.

It is obvious that points 2 and 3 describe a structural situation where a hydrogen atom bonded to one of the electronegative atoms [X in eq. (2.1)] is located approximately in the center between X and B, and the length of the hydrogen bond [H$\cdots$B in terms of eq. (2.1)] is remarkably smaller than a sum of the van der Waals radii of atoms H and B. For example, a typical length of hydrogen bonds in water molecules is 1.97 Å. It should be emphasized that the length of hydrogen bonds depends on *temperature* and *pressure*, which is obvious because interaction (2.1) is accompanied by reduced molecular volume. Finally, points 2 and 3 correspond to remarkable elongations in the initial proton-donating X–H bonds. These elongations, accompanied by the well-known *red shift* of X–H vibrational modes, also form the most important experimental criterion for the formation of hydrogen bonds.

A modern description of a conventional hydrogen bond as well as its older, more accurate definition are based on Bader's theory of atoms in molecules (AIM theory) [4]. Bader considers matter "a distribution of charge in real space of point-like nuclei embedded in the diffuse density of electron charge, $\rho(r)$. All the properties of matter are manifested in the charge distribution and the topology of the bonding between atoms. In a bound state, the nuclei of bonded atoms are linked by a line along which the electron density is a *maximum* with respect to any neighboring line." In other words, this theory suggests a self-consistent approach that is independent of crystallographic, spectroscopic, and other physic-ochemical experimental data and provides rigorous and unambiguous criteria for the determination of which atoms are chemically bonded in a molecule.

AIM theory takes electron density as a starting point. In this context, the electron density function $\rho$ represents a three-dimensional function which can be defined as follows: $\rho(r) \, dr$ is the probability of finding an electron in a small volume element $dr$ at some point in space characterized by the distance $r$. Bader emphasizes that the electronic density is a real object that can be investigated by computing.

The key to investigating the topology of the electron density $\rho$ is the gradient vector $\nabla\rho$, which is perpendicular to a constant electron density surface and points in the direction of steepest ascent. Then, a sequence of infinitesimal gradient vectors corresponds to a *gradient path*. Since gradient vectors are directed, gradient paths also have a direction: They can go uphill or downhill.

Typically, gradient paths are directed to a point in space called an *attractor*. It is obvious that gradient paths should be characterized by an endpoint and a starting point, which can be infinity or a special point in the molecule. All nuclei represent attractors, and the set of gradient paths is called an *atomic basin*, $\Omega$. This is one of the cornerstones of AIM theory because the atomic basin corresponds to the portion of space allocated to an atom, where properties can be integrated to give atomic properties. For example, integration of the $\rho$ function yields the atom's population.

The second very important point in AIM theory is its definition of a chemical bond, which in the context of gradient paths, is straightforward. In fact, some gradient paths do not start from infinity but from a special point, the *bond critical point*, located between two nuclei.

Bond critical points represent extremes of electronic density. For this reason, these points are located in space where the gradient vector $\nabla\rho$ vanishes. Then the two gradient paths, each of which starts at the bond critical point and ends at a nucleus, will be the atomic interaction line. When all the forces on all the nuclei vanish, the atomic interaction line represents a *bond path*. In practice, this line connects two nuclei which can consequently be called *bonded* [5]. In terms of topological analysis of the electron density, these critical points and paths of maximum electron density (atomic interaction lines) yield a *molecular graph*, which is a good representation of the bonding interactions.

Since all the parameters describing the electron density and the computing methods will be illustrated and discussed thoroughly in the context of dihydrogen bonds, in this section we are limited by enumeration of the AIM criteria that permit formulation of the interaction that establishes a classical hydrogen bond. These criteria are based on probing (1) the charge density, $\rho_C$, and the Laplacian of the charge density, $\nabla^2 \rho_C$, determined at bond critical points; (2) the topology of bond paths between a hydrogen atom and a hydrogen bond acceptor; (3) the mutual penetration of a hydrogen atom and an interacting bond; (4) the loss of charge and energetic destabilization of a hydrogen atom and the total charge transferred; and (5) the dipole moment enhancement. Using these modern terms summarized by Popelier [5], the interaction in eq. (2.1) can be formulated as a hydrogen bond under the following conditions:

1. The $\rho_C$ values are small and the $\nabla^2 \rho_C$ values are small and positive, corresponding to *closed-shell interactions* (in this context, the van der Waals complexes would have the smallest $\rho_C$ values).

2. A topological analysis of the bond paths between a hydrogen atom and a hydrogen bond acceptor actually establishes the existence of a bonding interaction.

3. The mutual penetrations of a hydrogen atom and a hydrogen bond acceptor are *positive*. The levels of penetration can be estimated by comparison of the nonbonded radii $r_H^0$ and $r_B^0$ in terms of eq. (2.1) (the *nonbonded radius* is the distance from a nucleus to a 0.001-au charge density contour in the direction of the hydrogen bond) and the corresponding bonded radii $r_H$ and $r_B$ or bond critical point radii. When all the penetrations are *positive*, the interactions can be formulated as hydrogen bonds.

4. Computations show the loss of charge in the hydrogen atom involved in the bond. This loss, $\Delta N$, computed by subtracting the electron population of the hydrogen atom in the free monomer from the corresponding hydrogen atom in the hydrogen-bonded complex, should be *negative*.

5. The hydrogen atom in the hydrogen-bonded complex is *destabilized*, and the destabilization, defined as the difference in total atomic energy between the hydrogen-bonded complex and the monomer, $\Delta E$, is *positive*.

6. The computations demonstrate that the dipole moment of the hydrogen-bonded complex is *larger* than the vector sum of the dipole moments of the monomers.

As has been demonstrated by Grabowski [6], all of these criteria, connected with the topology of the electron density, are suitable for the simplest hydrogen-bonded systems. Table 2.1 lists the geometrical, topological, and energy parameters of these systems computed at the MP2/6-311++G** level. As shown, the H$\cdots$B bond lengths, computed as 1.70 to 1.97 Å, are significantly smaller that the sum of the corresponding van der Waals radii. The $\rho_C$ values in the H$\cdots$B directions are low, and the $\nabla^2 \rho_C$ parameters take small positive values. In contrast, the individual H–F molecule shows a large electron density $\rho_C$ (0.371 au) and a very large *negative* Laplacian (−2.839 au). It is also clearly seen that the hydrogen bonding leads to an elongation in the H–F bond, which is accompanied by decreasing the $\rho_C$HF and $\nabla^2 \rho_C$HF parameters. In addition, it is easily seen that the topological, geometrical, and energy parameters are mutually correlated.

Criterion 1 can also be well illustrated by its application to the classical intramolecular hydrogen bonds O–H$\cdots$O in 2-acetyl-1,8-dihydroxy-3,6-dimethylnaphthalene (Structure 2.1). Theoretical and experimental data reported

**Structure 2.1** Intramolecular hydrogen bonds, O–H$\cdots$O, formed in a molecule of 2-acetyl-1,8-dihydroxy-3,6-dimethylnaphthalene.

**TABLE 2.1. Geometrical Parameters, Bonding Energies, and Topological Analysis of Electronic Density in the Framework of AIM Theory[a]**

| Complex | $r$(H–X) (Å) | $r$(H$\cdots$B) (Å) | $-\Delta E$ (kcal/mol) | $\rho_C$HX (au) | $\nabla^2\rho_C$HX (au) | $\rho_C$HB (au) | $\nabla^2\rho_C$HB (au) |
|---|---|---|---|---|---|---|---|
| HF$\cdots$H$_2$O[b] | 0.931 | 1.730 | 7.54 | 0.347 | $-2.652$ | 0.037 | 0.142 |
| HF$\cdots$NH$_3$ | 0.948 | 1.703 | 11.18 | 0,325 | $-2.365$ | 0.050 | 0.120 |
| HOH$\cdots$NH$_3$ | 0.972 | 1.974 | 5.77 | 0.348 | $-2.450$ | 0.028 | 0.085 |

[a]Obtained for hydrogen-bonded complexes X–H$\cdots$B (X–H is a proton donor site) at the MP2/6-311++G** level.
[b]For the individual HF molecule the $r$(H–X), $\rho_C$HX, and $\nabla^2\rho_C$HX values are calculated as 0.916 Å, 0.371 au, and $-2.839$ au, respectively.

**TABLE 2.2. Electronic Density Parameter at Bond Critical Points for Intermolecular Hydrogen Bonding in 2-Acetyl-1,8-dihydroxy-3, 6-dimethylnaphthalene[a]**

| Parameter | O(3)–H(1) | H(1)–O(2) | O(2)–H(2) | H(2)–O(1) |
|---|---|---|---|---|
| $\rho_C$ (au) | 0.264 | 0.112 | 0.048 | 0.337 |
| $\nabla^2\rho_C$ (au) | $-1.171$ | 0.092 | 0.144 | $-1.972$ |

[a]Computed at the B3LYP/6-31++G(d,p) level.

by Sorenson and co-workers [7] have shown two different intramolecular hydrogen bonds in this molecule. One of them, H(1)–O(2), is short and exhibits a very high bonding energy of $-22.1$ kcal/mol. The second bond, O(2)–H(2), is significantly weaker, with an energy gain of only $-9.5$ kcal/mol. The topological analysis of the electron density is in good agreement with the foregoing results. As follows from the data in Table 2.2, both hydrogen bonds exhibit $\rho_C$ values that are smaller than those of covalent H–O bonds. The $\nabla^2\rho_C$ values on the hydrogen bonds are positive, whereas they take negative values on covalent bonds. It is remarkable that the stronger hydrogen bond shows the higher electron density.

The above-mentioned topological criteria are very useful since the geometrical criteria are often not sufficient to establish whether or not hydrogen bonding exists. As we will see below, these theoretical criteria are universal and quite suitable for nonconventional hydrogen bonds as well as for dihydrogen bonding.

### 2.1.1. Energy and Geometry of Conventional Hydrogen Bonds

Energies of formation for conventional hydrogen bonds depend on the strength of a proton donor, X–H, and a proton acceptor, B, and also on the polarity of the environment, the temperature, and the pressure [8], varying from 1–2 to 40 kcal/mol for ionic hydrogen bonds (see below). It is obvious that 40 kcal/mol, measured, for example, in the ion $[HF_2]^-$, does not allow us to discriminate such hydrogen bonds and covalent chemical bonds. Generally, however, typical values of interaction energies for the hydrogen bonds O–H$\cdots$N, O–H$\cdots$O, O–H$\cdots$N, and N–H$\cdots$O are measured as $-7$, $-5$, $-3$, and $-2$ kcal/mol, respectively.

In terms of the bonding energy, conventional hydrogen bonds can be classified as *weak* when energies are between $-2.4$ and $-12$ kcal/mol, *strong* when energies are between $-12$ and $-24$ kcal/mol, and *very strong* when energies become higher than 24 kcal/mol. This classification, suggested by Hibbert and Emsley [9], seems very subjective. For example, Alkorta and co-workers [3] prefer a classification in which hydrogen bonds with energies of less than $-5$ kcal/mol are considered weak, bonds with energies between $-5$ and $-10$ kcal/mol are defined as medium, and strong and very strong hydrogen bonds correspond to energies that are larger than $-10$ kcal/mol. The latter, more reasonable, classification is used in this book.

The nature of hydrogen bonding can be well rationalized in the framework of modern theoretical calculations that provide estimates of the total energy of hydrogen-bonded complexes as the sum of separate terms. These terms can be obtained in the limits of an analysis of the energy decomposition suggested by Kitaura and Mokomura [10]. Here the separate contributions are associated with real physical effects such as repulsion, $E_{exrep}$; electrostatic interaction, $E_{elec}$; polarization, $E_{pol}$; and charge transfer, $E_{cht}$. Table 2.3 illustrates the results of

**TABLE 2.3. Energy Contributions to the Total Bonding Energy (kcal/mol) for Classical Hydrogen-Bonded Complexes Formed by an HNO Molecule[a]**

| Complex | $E_{elec}$ | $E_{exrep}$ | $E_{pol}$ | $E_{cht}$ | $E_{tot}$ |
|---------|-----------|-------------|-----------|-----------|-----------|
| HNO$\cdots$HF | $-6.22$ | 3.76 | $-0.77$ | $-1.74$ | $-4.94$ |
| HNO$\cdots$HOH | $-3.06$ | 1.69 | $-0.26$ | $-1.02$ | $-2.66$ |

[a] $E_{elec}$, electrostatic; $E_{exrep}$, exchange; $E_{pol}$, polarization; and $E_{cht}$, charge transfer. Calculated at the HF/6-31G*level.

**TABLE 2.4. Decomposition of Total Interaction Energies (kcal/mol) for Nonconventional Hydrogen-Bonded Complexes**[a]

| Energy[b] | Formaldehyde–CHCl$_3$ | Acetone–CHCl$_3$ | Benzene–Formaldehyde | 1,1-Dichloroethane–Acetone |
|---|---|---|---|---|
| $E_{elec}$ | −4.34 | −5.48 | −0.68 | −3.82 |
| $E_{exrep}$ | +2.32 | +3.05 | +0.28 | +1.7 |
| $E_{pol}$ | −0.43 | −0.65 | −0.11 | −0.43 |
| $E_{cht}$[c] | −0.33 | −0.53 | −0.047 | −0.25 |
| $E_{tot}$ | −2.79 | −3.61 | −0.56 | −2.81 |

[a]Performed at the restricted HF/6-31G** level.
[b]$E_{elec}$, energy of electrostatic interaction; $E_{exrep}$, energy of repulsion; $E_{pol}$, energy of polarization; $E_{cht}$, energy of charge transfer; $E_{tot}$, total energy.
[c]The basis set superposition error [13] has been eliminated from the calculation of charge transfer energies.

calculations at the HF/6-31G* level carried out by Orlova and Scheiner for two classical hydrogen-bonded complexes [11]. Koch and Popelier [12] have used the same approach and computed at the HF/6-31G* level the nonconventional hydrogen-bonded complexes formaldehyde–CHCl$_3$, acetone–CHCl$_3$, benzene–formaldehyde, and acetone/1,1-dichloroethane (Table 2.4). It follows from the calculations that the magnitudes of the separate contributions and their ratios and of the total bonding energies depend strongly on the nature of proton donors and proton acceptors. For example, the total bonding energy in Tables 2.3 and 2.4 changes between −2.6 and −4.9 kcal/mol or from −0.56 to −3.61 kcal/mol, respectively. It should be noted that energies obtained by different methods cannot be compared. However, it is a very important fact that two different methods applied to the various types of complexes give the same result: The electrostatic component is strongly dominant. This is typical of hydrogen bonds.

Since the main contribution to the total energy of hydrogen-bonded complexes is provided by electrostatic interaction, the presence of ionic charges will enhance the hydrogen-bonding strength. This is actually observed in *charge-assisted* or *ionic hydrogen-bonded complexes*, $X–H^+\cdots B$ or $X–H\cdots^- B$, where due to the additional polarization, the acidity of a proton-donor component and the basicity of a proton-acceptor site increase.

Numerous examples of ionic hydrogen bonds (e.g., in ref. 14) are addressed complexes that are formed as

$$BH^+ + B' \rightleftharpoons BH^+ \cdots B' \qquad (2.2)$$

where the hydrogen donor, the ion $BH^+$, is a protonated base and the hydrogen acceptor site is a regular base or as:

$$X^- + HX \rightleftharpoons X^- \cdots H–X \qquad (2.3)$$

where $X^-$ is a deprotonated Brønsted acid. Energies of formation for such hydrogen bonds cover a region between 5 and 35–40 kcal/mol. These interactions play an important role, for example, in ionic clusters and nucleation, and in the structures of ionic crystals, surfaces, silicates, and the like. They are also important in bioenergetics, including protein folding, enzyme active centers, formation of membranes, and proton transport [14]. The formation of ionic hydrogen bonds (2.2) is accompanied by a partial proton transfer from the proton donor to the proton acceptor. As result, the hydrogen atom becomes more positive and the heteroatoms in the proton-donor component and proton-accepting site take a more negative charge. Similar features can be found in anionic hydrogen bonds (2.3). It should be pointed out that the formation of ionic hydrogen bonds requires significant entropy changes ($\Delta S° = -20$ to $-60$ eu).

In the context of geometry, conventional hydrogen bonds can be classified as *linear* or *bifurcated*. Generally, conventional *intermolecular* hydrogen bonds, formed by a single donor–acceptor pair (e.g., O–H$\cdots$O or N–H$\cdots$O), are linear or very close to linearity. The reason for this effect can be found by quantum-chemical calculations, where electrostatic interaction, providing the main contribution to the total energy of the hydrogen bond, is reduced by 10% if the geometry of the X–H$\cdots$B fragment deviates from linearity by 20°. That is why 90% of the hydrogen bonds N–H$\cdots$O in protein molecules, for example, are characterized by a bond angle between 140 and 180°. Statistical data show that most bond angles average about 158° [15].

Remarkable deviations from the linear geometry are observed in the bifurcated hydrogen bonds that are formed by *one* proton-donor component and a *double* proton-acceptor group, as shown in Structure 2.2. Similarly, hydrogen bonds can be nonlinear if they are *intramolecular*, as in Structure 2.3, where the molecular geometry itself dictates the geometry of the bonds. It is worth mentioning that such intramolecular hydrogen bonds (called *resonance-assisted hydrogen bonds*) are particularly strong due to additional stabilization by the π-electronic system.

**Structure 2.2**

**Structure 2.3**

### 2.1.2. Cooperative and Anticooperative Energy Effects in Systems with Classical Hydrogen Bonds

Cooperative energy effects are particularly important when hydrogen-bonded complexes form cluster structures. For example, strong hydrogen bonds play a dominant role in enzyme catalysis [16], where they form *charge relay chains*. Such chains consist of a sequence of linear hydrogen-bonded molecules having mobile protons that can be transferred. These chains can be observed experimentally by low-temperature nuclear magnetic resonance (NMR) in solutions of acetic acid: for example, in $CDClF_2$–$CDF_3$, providing deep cooling [17].

Quantum-chemical calculations performed for dimers, trimers, and more complicated self-associates of simple molecules (e.g., H–Hal, $H_2O$, and HCN), have revealed that the hydrogen-bonding energies in the linear associates

$$\cdots X\text{–}H \cdots X\text{–}H \cdots X\text{–}H \cdots \qquad (2.4)$$

are remarkably higher than in the dimers, due to mutual polarization of the bonds. This cooperative effect increases with the chain length of the associates. For example, Table 2.5 illustrates calculations at the self-consistent field (SCF) level performed for the association of the HF molecule to give dimer, trimer, and tetramer clusters [18]. The data show clearly that the energy stability of the trimer formed by the HF molecule is very close to that of the dimer molecule if the energy is calculated on a per-hydrogen-bond basis. At the same time, the energy of the tetramer shows a sharp increase. Thus, the HF cluster growth is not smooth energetically with respect to the number of hydrogen bonds, illustrating the cooperativity of linear hydrogen bonds. Similar cooperative effects have been obtained in calculations of water clusters [19] and associated alcohols, amines, and amides.

In contrast to those noted above, theoretical investigations of branched complexes in which two or more hydrogen bonds are formed by one proton-acceptor group, as in Structure 2.4, predict an inverse effect. In this case, mutual polarization weakens the hydrogen bonds, leading to anticooperative effects. The various factors governing the influence of the first hydrogen bond on the formation of

**TABLE 2.5. Total and Relative Energies of Hydrogen-Bonded Clusters $(HF)_n^a$**

| System | HF | $(HF)_2$ | $(HF)_3$ | $(HF)_4$ |
|---|---|---|---|---|
| $E$ (au) | −100.06986 | −200.12828 | −300.20317 | −400.28170 |
| $\Delta E$ (kcal/mol) | 0 | 4.12 | 12.92 | 34.01 |

[a]Calculated at the SCF level.

$$(X-H\cdots X\cdots H-X)$$
$$|$$
$$H$$

**Structure 2.4**

the second by the same molecule or ion are described in the experimental work of Huyskens [20].

### 2.1.3. Dynamics of Classical Hydrogen Bonds

Since the strength of hydrogen bonds (typically, between 5 and 10 kcal/mol) is significantly smaller than that of covalent bonds ($10^2$ kcal/mol), molecular systems formed via hydrogen bonds show structural flexibility and fluctuation. These dynamics are critically important for chemical and biological catalysis and are manifested as fast vibrational motions of the hydrogen-bonded groups, breaking and formation of hydrogen bonds, and fast transfer of hydrogen atoms and protons. These motions cover a large time scale. Some of them are ultrafast and can be characterized by lifetimes of from 10 fs to picoseconds [21]. Such motions can be characterized by vibrational and ultrafast nonlinear vibrational spectra, which are very sensitive to hydrogen bonding (Figure 2.1). As shown, a low-concentration $C_2Cl_4$ solution of phenol (used to avoid self-association) shows

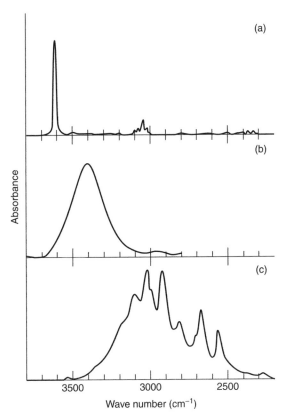

**Figure 2.1**  IR spectra in the $\nu$(OH) region recorded for (a) phenol in $C_2Cl_4$, (b) HOD in $D_2O$, and (c) $CD_3COOH$ in $CCl_4$. (Reproduced with permission from ref. 21.)

a quite narrow ν(OH) band. In contrast, the red-shifted band corresponding to weak hydrogen bonding in water is strongly *broadened*. In the case of strong hydrogen bonds ($CH_3COOH$), such broadening can be accompanied even by the appearance of new modes in Figure 2.1(c). Since similar effects observed in IR spectra are also typical of dihydrogen-bonded systems, it should be noted that the nature of the broadening is connected by anharmonic coupling between the high-frequency stretching motion and low-frequency hydrogen bond modes. The potential of the A−H stretching mode, illustrating enhanced anharmonicity, results from hydrogen bonding (Figure 2.2). In addition, this figure explains why the hydrogen bond length depends on the temperature. For detailed studies of the microscopic dynamics in hydrogen-bonded complexes, readers are referred to a recent review [21].

The dynamics in hydrogen-bonded complexes connected with their breaking and formation and also with proton transfer is of greatest interest in the context of comparing classical hydrogen bonds and dihydrogen-bonded systems. It is obvious that a solvent that acts as a strong base can blockade hydrogen bond formation. In liquids, where hydrogen bonds can produce structured networks, the dynamics are particularly interesting and complicated. The latter is clearly illustrated by liquid water, which is probably the most popular example of such systems.

Water molecules can yield from zero to four hydrogen bonds, which are constantly transforming: Long, weak hydrogen bonds become short, strong bonds, and vice versa. These hydrogen bond dynamics can be characterized by structural relaxation times $\tau_R$ and average lifetimes of hydrogen bonds $\tau_{HB}$. The average $\tau_{HB}$ times in liquid water are short (∼1 ps) and follow a simple Arrhenius dependence with an energy of 2.4 kcal/mol. This value is associated with the energy that is required for hydrogen bond breaks via librational motions, called "fast" motions. Diffusive motions, responsible for "slow" dynamics with times $\tau_R$, are more complex and follow a power law. Moreover, the dynamic behavior of the

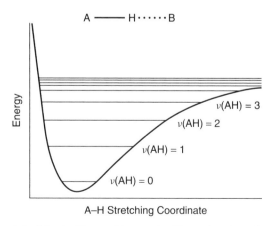

**Figure 2.2** Potential of the A−H stretching mode, illustrating the enhanced anharmonicity caused by formation of an A−H···B bond.

hydrogen bond network in water is different in different interfaces. For example, the dynamics of breaking and forming hydrogen bonds in the air–water interface is faster than that in bulk water, due to more rapid translational diffusion in the interface. This is in contrast to what happens on the surface of a protein [22]. Finally, it is worth mentioning that these processes are constant objects of theoretical [23] and experimental studies, implying the use of dielectric relaxation [24] or ultrafast infrared spectroscopic methods [25].

Generally, the formation of hydrogen-bonded complexes in solution is regarded as a very fast and diffusion-controlled process that has the energy necessary for reorganization of molecular environments. However, the formation of intramolecular hydrogen bonds can be limited kinetically by slower motions. A good example is the fluctuating hydrogen bonding in 2,6-dihydroxybenzoic acid depicted in Scheme 2.1. Here, two nonequivalent OH groups can be resolved in $^1$H NMR spectra at low temperatures, as shown in Figure 2.3. These OH resonances undergo a typical temperature evolution when heating leads to a single average line. Golubev and Denisov [26] have emphasized that the splitting of OH resonances at the lowest temperatures is connected with a slowing down of the internal rotation of the COOH group around a single C–C bond.

Jumps of a proton along the hydrogen bond represent another type of dynamics observed in hydrogen-bonded complexes. Mechanistically, this process is simplest for intramolecular hydrogen bonds. The fast enol–enolic equilibrium shown in Scheme 2.2 illustrates an intramolecular proton-jumping system [27]. Here, substituent X dictates the equilibrium constant as well as the rate of proton transfer. It should be noted that such proton jumps can be stopped on the $^1$H NMR time scale only at very low temperatures.

Fast enol–enolic equilibrium is of great interest when the proton-jumping system is symmetric (Scheme 2.3). In these molecular systems, the consequent synergistic reinforcement of hydrogen bonding and π-delocalization produce the strong intramolecular resonance-assisted hydrogen bonds O–H···O. Due to the influence of the substitutions, increasing delocalization can transform the hydrogen bond from displaying asymmetric interaction (a double well) to displaying symmetric interaction (a single well) with the shortest O···O distance [28].

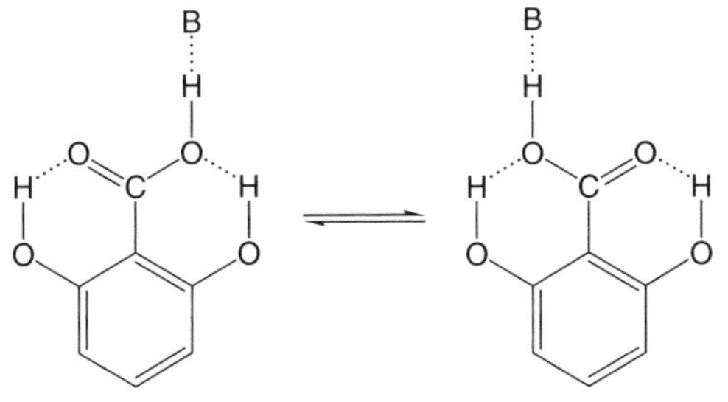

**Scheme 2.1** Schematic representation of intramolecular molecular dynamics in hydrogen-bonded 2,6-dihydroxybenzoic acid.

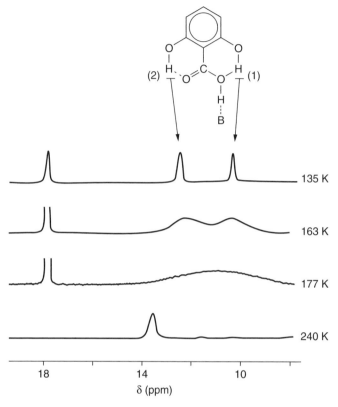

**Figure 2.3** Variable-temperature $^1$H NMR spectra recorded in a hexamethyl phosphoric triamide solution of 2,6-dihydroxybenzoic acid at 135, 163, 177, and 240 K. (Reproduced with permission from ref. 26.)

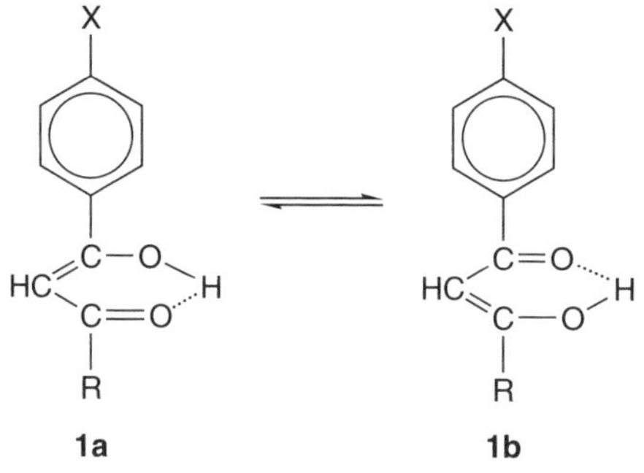

**Scheme 2.2** Schematic representation of a proton-jumping molecular system with fast enol–enolic equilibrium between structures **1a** and **1b**.

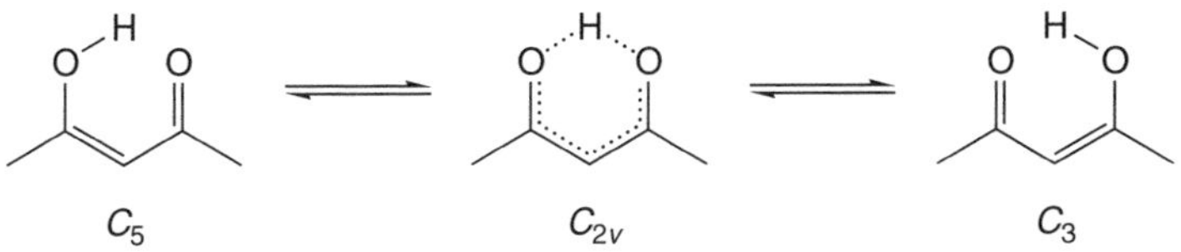

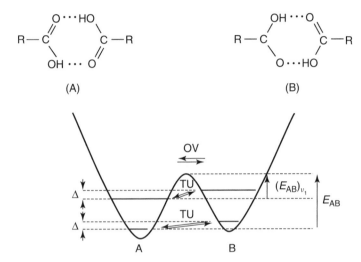

**Scheme 2.3**  Symmetrical enol–enolic system in which the structure with $C_{2v}$ symmetry is a transition state of the proton transfer.

**Figure 2.4**  Carboxylic acid dimer in the potential energy minima with local vibrational states. OV represents the correlation time for a thermally activated proton transfer, and TU, the correlation time for tunneling transfer. (Reproduced with permission from ref. 29.)

In reality, even simple intramolecular proton transfers can be complicated by proton tunneling effects. Figure 2.4 shows the potential energy minima with local vibrational states that correspond to dimers A and B of a carboxylic acid. Latanowicz and Medycki [29] have demonstrated that an elementary act of proton transfer consists of two independent constituent motions: classical hopping over the energy barrier, and incoherent tunneling through the barrier. Since both motions take place between the same potential minima, the geometry of these motions is identical.

Commonly, intermolecular proton transfer along a hydrogen bond, B, can be represented as the formation of a new hydrogen-bonded complex, C:

$$X{-}^{\delta+}H{+}\ :B \rightleftharpoons X{-}H\cdots :B \rightleftharpoons X^{\delta-}\cdots H{-}B^{\delta+} \rightleftharpoons X^{-}+H{-}B^{+}$$

$$\quad\quad A \quad\quad\quad\quad B \quad\quad\quad\quad C \quad\quad\quad D \quad\quad\quad\quad (2.5)$$

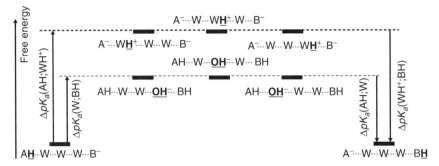

**Figure 2.5**   Relative energetics of the Grotthuss and proton-hole-transfer mechanisms as a function of the p$K_a$ values. (Reproduced with permission from ref. 30.)

which is a hydrogen-bond-stabilized ion pair that exists as an intermediate. In this equilibrium, proton transfer is completed when intermediate C is transformed into solvate-separated ions D. Here the proton transfer along the hydrogen bond occurs faster when the hydrogen bond is stronger (shorter). It should be noted that equilibrium (2.5) corresponds to *short-range intermolecular proton transfer*. Such proton transfers are mechanistically relatively simple. By contrast, the detailed mechanism of *long-range proton transfer processes*, particularly in solutions of biological systems, is significantly more complex. In these systems, sequential proton hops from the initial proton donor molecules to proton acceptors are mediated by water molecules or other groups that are capable of ionization. This can happen in two ways (Figure 2.5): the Grotthuss and proton-hole-transfer mechanisms, which depend strongly on the $pK_a$ values of donor–acceptors AH and BH, water W, and the hydronium ion WH$^+$. As shown, some combinations of p$K$ values (when both AH and BH components have p$K$ values higher than 7) lead to a situation in which the proton-hole-transfer pathway can become more energetically favorable [30].

## 2.2. NONCONVENTIONAL HYDROGEN BONDS AS A PART OF HYDROGEN-BONDED SYSTEMS: DEFINITION AND CLASSIFICATION

Hydrogen bonds [3] can be classified as nonconventional (or nonclassical), when they are formed with the participation of nonconventional proton donors and/or unconventional proton acceptors:

$$X–H \cdots B \tag{2.6}$$

where X is a carbon atom and B is isocyanides (R$-$N$\equiv$C:), carbenes (C:), carbanions (C$^-$), or $\pi$-electronic systems.

Among $\pi$-electronic systems acting as proton acceptors, the gold ethyne and benzene complexes are best known. Some of them have been characterized by x-ray crystallography: for example, the T-shaped $Cl_3C-H \cdots \pi(C\equiv C)$ hydrogen-bonded system [3]. The other complexes became objects of theoretical studies. According to calculations carried out at the MP2/6-311++G** and B3LYP/6-311++G** levels, isocyanides can act as proton acceptors to form hydrogen bonds:

$$H-N \equiv C : +HX \rightleftharpoons H-N \equiv C : \cdots H-X \rightleftharpoons [H-N \equiv C-H]^+ \qquad (2.7)$$

with different proton donors, such as HF, HOH, $HNH_2$, and HCN. Figure 2.6 shows a relief map of the electronic density, obtained in the framework of AIM theory, for an HF molecule interacting with $C\equiv N-H$. Similar to classical hydrogen bonds, the nonconventional hydrogen-bonded complex is linear and manifests lengthening of the H–N bond and shortening of the $N\equiv C$ bond.

Hydrogen-bonded complexes related to systems in eq. (2.7) have been found by the calculation of molecules $H_2C:$, $H_2Si:$, CO, and $F_2C:$ in the presence of various proton donors. Interaction energies have been calculated between $-2$ and $-22$ kcal/mol. As indicated in Table 2.6, the interaction energy depends greatly on the strength of proton donors. Thus, similar to classical hydrogen bonds, nonconventional hydrogen-bonded systems can be weak, medium, or even strong.

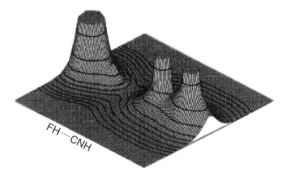

**Figure 2.6** Relief map of the electronic density corresponding to the complex FH$\cdots$:CNH obtained in the framework of the AIM methodology. (Reproduced with permission from ref. 3.)

**TABLE 2.6. Bonding Energies (kcal/mol) of Hydrogen-Bonded Complexes Formed by Nonconventional Proton-Acceptor Sites**[a]

|        | CNH   | CO    | CS    | $CH_2$ |
|--------|-------|-------|-------|--------|
| FH     | 5.763 | 3.356 | 7.253 | 12.070 |
| NCH    | 3.891 | 1.253 | 3.450 | 5.458  |
| $H_2O$ | 2.849 | 1.324 | 3.254 | 5.921  |
| HC≡CH  | 1.813 | 0.568 | 1.551 | 2.681  |

[a]Calculated at the B3LYP/6-311++G**level.

However, calculations related to the gas phase usually overestimate bonding energies compared to those expected in solution.

Hydrogen bonds $X-H \cdots M$, where d-electrons in metal atoms play the role of proton-acceptor sites, can also be classified as nonconventional hydrogen bonds. This type of hydrogen bond has recently been reviewed by Shubina and co-workers [31], which prompts us to discuss briefly some examples and main conclusions. As in the case of classical hydrogen bonding, metal–proton bonds can be intramolecular and intermolecular. Metallocenylcarbinoles in Structure 2.5, where M=Fe, Ru, Os, illustrate intramolecular interactions $M \cdots H$ with distances smaller than the sum of the van der Waals radii of H and M. It is worth noting that experiments have indicated that the molecules shown in Structure 2.5 do not have a preferable conformation that could dictate the proximity of the OH group to the metal atom, for example, due to steric factors.

The IR bands of metal-coordinated O–H groups are easily observed in dilute solutions of hydrogen-bonded complexes. $\Delta v(OH)$ shifts measured in inert solvents can be used conveniently as a measure of their relative strength. These shifts show that the bonding energy depends strongly on the nature of the metal center. For example, the strength of intramolecular $M \cdots H$ bonding in the complexes in Structure 2.5 and in isostructural transition metal carbinols increase in the order Mn < Re and Fe < Ru < Os.

In contrast to solutions, because of the proximity of molecules in the solid state, the existence of $M \cdots H$ intramolecular hydrogen bonding or its absence is controlled by competition between intermolecular classical bonds $O-H \cdots O$ and intramolecular interactions $O-H \cdots M$. It has been found that $M \cdots H$ bonding is more effective when the proton-accepting strength of the metal atom increases.

Intermolecular hydrogen bonds $M \cdots H$ have been found in solutions containing organic acids (or acidic alcohols) and metallocenes, $Cp_2M$; decamethylmetallocenes, $Cp_2^*M$; or half-sandwich complexes, $CpML_2$. Again, hydrogen-bonded complexes found by IR spectra in $v(OH)$ regions and $\Delta v(OH)$ red shifts, varying from 200 to 500 $cm^{-1}$, have been used to determine the relative strength of nonconventional hydrogen bonding. As earlier, the strength of intermolecular proton–metal bonding depends on the metal atom, increasing in the order Co < Rh < Ir, Ru < Os, and Mo < W.

Studies of organometallic compounds show a complete analogy between hydrogen bonding with the participation of d-electrons in metal atoms and sp-electrons in heteroatoms in regular organic bases. For example, in the context

**Structure 2.5**

of vibrational modes, the isotopic ratio $\Delta\nu(OH\cdots M)/\Delta\nu(OD\cdots M)$ in organo-metallic hydrogen-bonded complexes is measured as 1.35, as in the case of regular hydrogen bonds. DFT/B3LYP and EHF calculations of the $M\cdots H$ bonded systems $Cp_2Os\cdots HOH$ have revealed that lone pairs of d-electrons behave as lone pairs in the oxygen atom [32]. The enthalpies of $M\cdots H$ bond formation, obtained from 3 to 7 kcal/mol, are very close to those measured for classical hydrogen bonds of medium strength. The solvent effects are almost the same: Increasing the polarity of the solvent leads to decreasing the energy of $M\cdots H$ bond formation. For example, the formation enthalpy, measured experimentally for the complex $Cp_2*Ir(CO)_2\cdots HOCH(CF_3)_2$, increases as

$$CH_2Cl_2(4.8\ \text{kcal/mol}) < CCl_4(5.7\ \text{kcal/mol}) < C_6H_{14}(6.1\ \text{kcal/mol}) \qquad (2.8)$$

The geometry of $M\cdots H-X$ bonds is again close to linearity according to theoretical studies of the $Cp_2Os\cdots HOH$ system. Neutron diffraction data collected for the $Et_3NH^+\cdots Co(CO)_4$ complex have determined the $Co\cdots H-N$ angle to be $180°$.

It is also remarkable that $M\cdots H-X$ systems show proton transfer which is mechanistically close to that of classical hydrogen bonds. Protonation of the metal atom occurs via $M\cdots H$ bonds and ion pairs (Scheme 2.4) lying as intermediates on the reaction coordinate. As we will see below, a very similar mechanism operates in the case of dihydrogen-bonded complexes.

An *inverse hydrogen bond* is formed when a hydrogen atom can give electrons while another, *nonhydrogen atom* is accepting them:

$$A-H^{-\delta\cdots+\delta}B \qquad (2.9)$$

For example, the hydrogen atoms of the strongly polarized bonds in hydrides LiH and $BeH_2$ or $BH_4^-$ can be electron donors, and the electron-deficient atoms Li, Be, or B can accept electrons to form inverse hydrogen-bonded complexes: $Li-H\cdots Li-H$, $H-Be-H\cdots Li-H$, and others [3]. Similar to classical hydrogen bonds, the electronic distribution in these inverse hydrogen bonds, analyzed in the framework of AIM theory, shows that the hydrogen atom is bound to both the electron donor and the electron acceptor by *closed-shell interactions*. In addition, the bond critical points correspond to all the characteristics associated

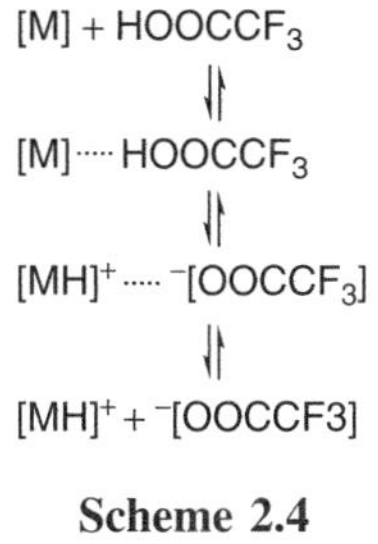

$$[M] + HOOCCF_3$$
$$\updownarrow$$
$$[M]\cdots HOOCCF_3$$
$$\updownarrow$$
$$[MH]^+\cdots {}^-[OOCCF_3]$$
$$\updownarrow$$
$$[MH]^+ + {}^-[OOCCF3]$$

**Scheme 2.4**

with hydrogen bonds: The $\rho_C$ values are low, and the Laplacian, $\nabla^2 \rho_C$, is positive. Finally, the bonding energies in inverse hydrogen bonds can be significant, reaching values between $-5$ and $-10$ kcal/mol. For example, in the LiH dimer, the bonding energy is as high as $-25$ kcal/mol.

## 2.3. DIFFERENCE BETWEEN HYDROGEN AND CHEMICAL BONDS

As we have shown above, hydrogen bonding is diverse energetically and geometrically. The bond lengths and dynamics of hydrogen-bonded complexes are also different. A hydrogen bridge can appear between proton acceptors and proton donors of a different nature, symbolizing classical, nonclassical, or inverse hydrogen bonds. However, despite this variety, experimental and theoretical studies demonstrate that there is no principal difference between these different types of hydrogen bonding, and in some sense their separation is somewhat classifying in character. In this section we show a general view that reveals the similarity between hydrogen bonds and covalent chemical bonds [33,34].

Electrostatic interactions between polarized proton donor X–H and acceptor B have been shown to provide the main contributions to the total bonding energy in an X–H$\cdots$B bridge. At the same time, many experimental observations agree regarding the presence of *covalency* in *hydrogen bonding*. One of the most important facts demonstrating the validity of this statement is the *scalar nuclear spin–spin coupling* through hydrogen bonds observed in NMR spectra (such coupling is provided by Fermi contact interaction, the meaning of which is explained below). For example, the remarkable spin–spin coupling constants are easily detected by NMR experiments with hydrogen-bonded clusters [F(HF)$_n$]$^-$ [J($^{19}$F–$^{19}$F)] and $^{15}$N-labeled hydrogen-bonded anions [C≡$^{15}$N$\cdots$H$\cdots$$^{15}$N≡C]$^-$ [33]. It is worth mentioning that the constants observed are well reproduced by high-level ab initio calculations. Experiments and theoretical treatments show that the coupling constant J(X–B) between the two heavy nuclei in a hydrogen-bonded fragment, X–H$\cdots$B, can be larger than the constants J(XH) and J(HB). In other words, the data illustrate *direct positive overlap of the atomic orbitals* of X, B, and H. It is remarkable that the J(X–B) constants exhibit a maximum when the proton is shifted from the X atom toward the B, to give the shortest X$\cdots$B distances in symmetric or quasisymmetric hydrogen-bonded systems.

These results can be interpreted successfully in terms of Pauling's valence bond order concept. In the framework of this model, a chemical bond between X and H in diatomic molecule XH or between H and B in a HB molecule can be characterized by empirical valence bond orders $P_{XH}$ and $P_{HB}$ decreasing exponentially with bond distance:

$$P_{XH} = \exp\left[\frac{-\left(r_{XH} - r_{XH}^0\right)}{b_{XH}}\right]$$

$$P_{HB} = \exp\left[\frac{-\left(r_{HB} - r_{HB}^0\right)}{b_{HB}}\right] \tag{2.10}$$

Here $r_{XH}^0$ and $r_{HB}^0$ are the equilibrium bond lengths and $b_{XH}$ and $b_{HB}$ are the parameters that describe the decreasing bond order. It is obvious that at $r_{HB} = r_{HB}^0$ or $r_{XH} = r_{XH}^0$, the $P_{XH}$ and $P_{HB}$ transform to unity. Then, going to the hydrogen bound to both X and B atoms in a hydrogen-bonded complex $X-H\cdots B$, its total valence remains equal to unity:

$$P_{XH} + P_{HB} = 1 \qquad (2.11)$$

This equation implies that the difference between the usual covalent bond and the hydrogen bond is not qualitative but has only a quantitative character: The hydrogen bond exhibits the smaller valence bond order. In these terms, for example, hydrogen bonds $H\cdots O$ in ice correspond to a covalent bonding of 5%.

Additional, particularly strong experimental evidence for the partly covalent nature of hydrogen bonds is that of a low-temperature x-ray and neutron diffraction study of benzoylacetone, where charge density analysis has shown that the hydrogen position is stabilized by electrostatic and covalent contributions at each side of the hydrogen atom [35]. In accordance with this experiment, Gilli and Gilli have introduced the *electrostatic-covalent hydrogen-bond model* [36], where weak hydrogen bonds are electrostatic interactions but become increasing covalent when their strength increases. The very strong hydrogen bonds are in principle three-center four-electron covalent bonds. In other words, there is no restricted border between hydrogen bonds and covalent bonds.

## 2.4. CONCLUDING REMARKS

1. Hydrogen bonds $X-H\cdots B$ are classified as classical (B = O, N, etc.), nonclassical or nonconventional (B = carbenes, carbanions, or $\pi$-electronic systems and metal atoms), and inverse $A-H^{-\delta\cdots+\delta}B$ (where the hydrogen atom provides electrons that another, nonhydrogen atom accepts).

2. Energetically, classical hydrogen bonds are classified as weak when their bonding energies are below $-5$ kcal/mol, medium with energies between $-5$ and $-10$ kcal/mol, and strong or very strong with energies above $-10$ kcal/mol. Charge-assisted hydrogen bonds are the strongest. Bonding energies are dominated by electrostatic contributions.

3. An interaction can be termed a hydrogen bond when it corresponds to modern criteria based on the topology of the electron density.

4. Conventional hydrogen bonds can be linear or bifurcated. Generally, intermolecular conventional hydrogen bonds formed by a single donor–acceptor pair are linear or very close to linearity. The geometry of intramolecular hydrogen bonds is dictated by the molecular geometry.

5. Hydrogen bonds can be cooperative or anticooperative based on their mutual polarization.

6. The difference between the usual covalent bond and a hydrogen bond is only quantitative in character: The hydrogen bond exhibits the smaller valence bond order.

7. Since the strength of hydrogen bonds is significantly smaller than that of covalent bonds, molecular systems formed via hydrogen bonds show structural flexibility and fluctuations. In other words, hydrogen bonds are highly dynamic.

8. Mechanistically, short-range intermolecular proton transfers are relatively simple. However, even for systems with intramolecular hydrogen bonds, proton transfers can be complicated by the presence of tunneling effects.

9. The mechanism of long-range proton transfer processes in solutions is complex because sequential proton hops from initial proton donors to proton acceptors are mediated by water (or solvent) molecules or other groups capable of ionization.

## REFERENCES

1. M. C. Etter, *Acc. Chem. Res.* (1990), **23**, 120.
2. G. R. Desiraju, *J. Chem. Soc. Dalton Trans.* (2000), 3745.
3. I. Alkorta, I. Rozas, J. Elguero, *Chem. Soc. Rev.* (1998), **27**, 163.
4. R. F. W. Bader, *Atoms in Molecules: A Quantum Theory*, Oxford University Press, New York (1990).
5. P. L. A. Popelier, *J. Phys. Chem. A* (1998), **102**, 1873.
6. S. J. Grabowski, *Chem. Phys. Lett.* (2001), **338**, 361.
7. J. Sorensen, H. F. Clausen, R. D. Poulsen, J. Overgaard, B. Schiott, *J. Phys. Chem. A* (2007), **111**>, 345.
8. G. A. Jeffrey, *An Introduction to Hydrogen Bonding*, Oxford University Press, Oxford (1997).
9. F. Hibbert, J. Emsley, *Adv. Phys. Org. Chem.* (1990), **26**, 255.
10. K. Kitaura, K. Mokomura, *Int. J. Quantum Chem.* (1976), **10**, 325.
11. G. Orlova, S. Scheiner, *J. Phys. Chem. A* (1998), **102**, 260.
12. U. Koch, P. L. A. Popelier, *J. Phys. Chem.* (1995), **99**, 9747.
13. S. F. Boys, F. Bernardi, *Mol. Phys.* (1970), **19**, 553.
14. M. Meot-Ner, *Chem. Rev.* (2005), **105**, 213.
15. E. Baker, R. Hubbard, *Prog. Biophys. Mol. Biol.* (1984), **44**, 97.
16. P. F. Cook, *Enzyme Mechanisms from Isotope Effects*, CRC Press, New York (1992).
17. N. S. Golubev, S. N. Smirnov, V. A. Gindin, G. S. Denisov, H. Benedict, H. H. Limbach, *J. Am. Chem. Soc.* (1994), **116**, 12055.
18. S. Y. Liu, D. W. Michael, C. E. Dykstra, J. M. Lisy, *J. Chem. Phys.* (1986), **84**, 5032.
19. R. Brakaspathy, S. Singh, *Chem. Phys. Lett.* (1986), **131**, 394.
20. P. L. Huyskens, *J. Am. Chem. Soc.* (1977), **99**, 2578.
21. E. T. Nibberring, T. Elsaesser, *Chem. Rev.* (2004), **104**, 1887.
22. P. L. E. Harder, B. J. Berne, *J. Phys. Chem. B* (2005), **82**, 2294.
23. F. W. Starr, J. K. Nielsen, H. E. Stantley, *Phys. Rev. Lett.* (1999), **82**, 2294.
24. J. B. Hasted, S. K. Husain, F. A. M. Frescura, J. R. Brich, *Chem. Phys. Lett.* (1985), **118**, 622.
25. C. P. Lawrence, J. L. Skinner, *J. Chem. Phys.* (2003), **118**, 264.

26. N. S. Golubev, G. S. Denisov, *J. Mol. Struct.* (1992) **270**, 263.

27. V. A. Gindin, I. A. Vhripun, B. A. Ershov, A. I. Koltsov, *Org. Magn. Reson.* (1972), **4**, 63.

28. R. Srinivasan, J. S. Feenstra, S. T. Park, S. Xu, A. H. Zewail, *J. Am. Chem. Soc.* (2004), **126**, 2266.

29. L. Latanowicz, W. Medycki, *J. Phys. Chem. A* (2007), **111**, 1351.

30. D. Riccardi, P. Konig, X. Prat-Resina, H. Yu, M. Elstner, T. Frauenheim, Q. Cui, *J. Am. Chem. Soc.* (2006), **128**, 16302.

31. E. S. Shubina, N. V. Belkova, L. M. Epstein, *J. Organomet. Chem.* (1997), **536–537**, 17.

32. G. Orlova, S. Scheiner, *Organometallics* (1998), **17**, 4362.

33. H. Benedict, I. G. Shenderovich, O. L. Malkina, V. I. Malkin, G. S. Denisov, N. S. Golubev, H. H. Limbach, *J. Am. Chem. Soc.* (2000), **122**, 1979.

34. P. Schuster, G. Zundel, C. Sandorfy, Eds., *The Hydrogen Bond*, North-Holland, Amsterdam (1976).

35. G. K. Madsen, B. B. Iversen, F. K. Larsen, M. Kapon, G. M. Reisner, F. H. Herbstein, *J. Am. Chem. Soc.* (1998), **120**, 10040.

36. G. Gilli, P. Gilli, *J. Mol. Struct.* (2000), **552**, 1.

# 3

# CONCEPT OF DIHYDROGEN BONDING

Following our general classification of hydrogen bonds, a negatively polarized hydrogen atom in $^{\delta-}H-Y$, where Y is a less electronegative element than H (e.g., B, Al, Re, K, Mg, etc.), could be considered an unusual or nonconventional proton acceptor. Then the interaction

$$X-H^{+\delta\cdots-\delta}H-Y \qquad (3.1)$$

could be formulated as a nonconventional hydrogen bond termed a *dihydrogen bond* by Wessel et al. [1]. What makes this bond so unusual is that the proton-accepting atom is also a hydrogen atom. It should be pointed out that the term *dihydrogen bond* should be distinguished from the term *dihydrogen ligand*, which corresponds to the binding of molecular hydrogen to a transition metal in a $^2\eta-H_2$ fashion. We also discuss the term *dihydrogen interaction* or H–H *bonding*, which describes the situation when two hydrogen atoms bind to each other in the absence of opposite charge polarizations (see below). Finally, we look at the formulation of the dihydrogen bond shown in Structure 3.1, where a σ B–H bond plays the role of an electron donor. Although the latter is probably more correct theoretically, in this book we use the formulation in eq. (3.1) as being the most common. Despite its unusual character, this type of proton–hydride interaction is now recognized as a common phenomenon in chemistry. For example, Alkorta and co-workers [2] believe that its spreading corresponds to Figure 3.1.

It is well known that among hydrogen-bonded complexes, the nonclassical π-hydrogen bonds are the weakest energetically. In fact, the stabilizing energy of π-hydrogen-bonded complexes is half of that typical of classical hydrogen bonds involving lone electron pairs. If these data are extrapolated straightforwardly to a σ chemical bond, they should be extremely weak hydrogen bond acceptors. Surprisingly, both experiments and theory have demonstrated that metal–hydrogen (M–H) and B–H σ-bonds act as proton acceptors with respect to acidic groups such as N–H or O–H, yielding medium or strong unconventional dihydrogen

---

*Dihydrogen Bonds: Principles, Experiments, and Applications*, By Vladimir I. Bakhmutov
Copyright © 2008 John Wiley & Sons, Inc.

**Structure 3.1**

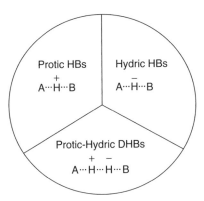

**Figure 3.1** Area occupied by dihydrogen bonds among other hydrogen-bonded complexes according to Alkorta et al. [2].

bonds that affect markedly even the physical properties of molecules. The latter follows from a logical comparison by Wessel et al. [1] of two molecules, $CH_3CH_3$ and $BH_3NH_3$, which are *isoelectronic* but whose physical properties differ strongly. The melting point is $-180°C$ for $CH_3CH_3$ and $+104°C$ for $BH_3NH_3$. Some part of this considerable difference could be attributed to the larger polarity of $BH_3NH_3$, which has a dipole moment of 5.2 D. However, the $CH_3F$ molecule, also being polar ($\mu = 1.8$ D), still shows a very low melting point, $-140°C$. Thus, even this simple consideration leads to the conclusion that there are other factors (e.g., dihydrogen bonding) that affect the physical properties of $BH_3-NH_3$, due to the formation of an extended network in the solid state. In other words, strength and directionality of dihydrogen bonds are comparable with those established for conventional hydrogen bonds. As we show below, dihydrogen bonds can actually affect the structure of compounds in solution and in the solid state and also the reactivity and even the selectivity of reactions. In addition, dihydrogen bonds are finding potential uses in catalysis and material chemistry. Finally, dihydrogen bonding can also be a good tool with which to build periodic supramolecules in crystal engineering [3].

## 3.1. GENERAL VIEW: FROM AN H₂ MOLECULE TO A DIHYDROGEN BOND VIA A DIHYDROGEN LIGAND

As mentioned in Chapter 2, the difference between a covalent chemical bond and a hydrogen bond is only quantitative, despite the dramatic difference in their energies: The hydrogen bond exhibits a smaller valence bond order. Then, following this logic, the $H_2$ molecule,

$$H\text{–}H \tag{3.2}$$

provides a well-known example of the strongest dihydrogen bonding. In our quest for general representations, let us look at what H–H bonding means in this case.

In terms of quantum chemistry, an H–H bond is formed when the electron wave functions of hydrogen atoms overlap. In turn, this overlap corresponds to an exchange interaction. In the usual case, the wave function describing the two electrons in a pair of atoms (and also their distribution) can be symmetric or anti-symmetric with respect to the exchange of identical electrons. As is clear from Figure 3.2, a symmetric wave function corresponds to a bonding configuration of electrons when the hydrogen atoms approach each other. In contrast, an anti-symmetric wave function produces an antibonding electron configuration. The electron density, expressed as the square of the magnitude of the wave function, $\psi_S^2$ and $\psi_A^2$ in Figure 3.2, shows that the symmetric function produces a high electron density in the area between two nuclei and thus leads to the appearance of a net attractive force between atoms. In the case of hydrogen atoms, the exchange interaction above produces a strong covalent bond with a dissociation energy of 4.52 eV at an H–H separation of 0.74 Å (Figure 3.3). The potential energy $U_A$ of the antibonding orbital shows clearly why a third hydrogen atom cannot be bonded to the two atoms in the hydrogen molecule. It should be pointed out that examination of the energy balance of formation of ions $H^+$ and $H^-$ and the attractive force between them leads to the energy required, which is always *positive* at any value of $H^+/H^-$ separation. In other words, there is no distance at which a net attractive interaction is present, and thus the H–H bond has no ionic character.

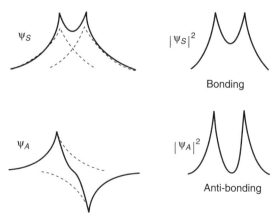

**Figure 3.2** Symmetric and antisymmetric wave functions, describing the two electrons in a pair of atoms and leading to bonding and antibonding electron configurations.

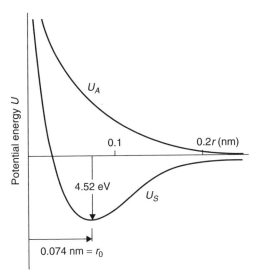

**Figure 3.3** Potential energy in the H$_2$ molecule corresponding to the bonding ($U_S$) and antibonding ($U_A$) orbitals.

In the framework of AIM theory, which well be used extensively to describe dihydrogen bonds, the situation above corresponds to the electron density distribution in the H$_2$ molecule, where the $\rho_C$ parameter takes the very large value 1.857 au and the Laplacian, $\nabla^2 \rho_C$, is strongly negative ($-34.15$ au). These data have been obtained by ab initio calculations at the MP2/6-31G* level [4].

Another type of dihydrogen bonding, one supported by a transition metal atom, has been known since 1984, when Kubas discovered that the hydrogen molecule can act as a ligand in transition metal complexes without cleavage [5]. The bonding interaction between a metal atom and hydrogen atoms has been rationalized in terms of a donation from the filled σ-bonding orbital of H$_2$ into an empty orbital on the metal that has σ symmetry. This interaction is balanced by back-donation from filled metal orbitals predominantly d in character to the σ*-orbital of an H$_2$ molecule. Figure 3.4 shows this type of bonding, in which this very delicate electron balance leads to a weaker H–H bond with H–H distances from 0.8 to 1.35 Å versus 0.74 Å in the hydrogen molecule [6].

Next, let us consider the dihydrogen ligand in the context of Bader's formalism (described in Chapter 2). With the help of DFT/B3LYP calculations, Maseras and co-workers [7] have analyzed topologically the electron density in the W and Ir dihydrogen complexes depicted in Figure 3.5. It is worth mentioning that the H–H bond lengths in these complexes are different and are 0.818 and 0.984 Å, respectively. In addition, they correspond closely to the various parameters that characterize the electron density (Table 3.1). Compared to the data obtained for the H$_2$ molecule, the data in the table illustrate the dramatic decrease in the $\rho_C$ and $\nabla^2 \rho_C$ parameters in the H–H directions when the free H$_2$ molecule becomes a dihydrogen ligand. Figure 3.6 shows the electron density map deduced for the

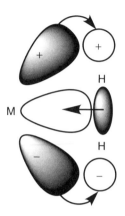

**Figure 3.4** $H_2$ binding to a transition metal atom in $^2\eta$ fashion. (Reproduced with permission from ref. 3.)

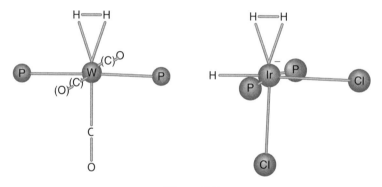

**Figure 3.5**

**TABLE 3.1. Structural and Electronic Properties of H–H and M–H Bonds in the W and Ir Dihydrogen Complexes Shown in Figure 3.5**

| Complex | Bond | Type | $\rho_C$ (au)[a] | $\nabla^2\rho_C$ (au)[b] | Bond Length (Å) |
|---------|------|------|------------------|--------------------------|-----------------|
| W | H–H | $(3, -1)$ | 0.226 | $-0.849$ | 0.818 |
|   | T–bond | $(3, -1)$ | 0.061 | $+0.308$ | |
| Ir | H–H | $(3, -1)$ | 0.164 | $-0.296$ | 0.984 |
|    | Ir–H | $(3, -1)$ | 0.127 | $+0.266$ | 1.651 |
|    |      | $(3, +1)$ | 0.119 | $+0.386$ | |

[a] 1 au $= 6.748$ e/Å$^3$.
[b] 1 au $= 24.10$ e/Å$^5$.

**Figure 3.6** Electron density contour map of an optimized W dihydrogen complex. The lines plotted correspond to the values 0.0001, 0.001, 0.01, 0.02, 0.04, 0.06, 0.08, 0.10, 0.13, 0.16, 0.19, 0.22, 0.26, 0.30, 0.35, 0.40, 0.70, and 1.00 au. (Reproduced with permission from ref. 7.)

W dihydrogen complex. This map reveals a very clear bond path connecting the two hydrogen atoms. As shown, the metal atom is not connected separately to each hydrogen atom but is connected to the bond critical point, reflecting a T-shaped bond structure. Therefore, a $(3, -1)$ critical point links a maximum charge density with a bond path. The positive Laplacian in Table 3.1 corresponds to closed-shell interactions. It should be emphasized again that the $\nabla^2 \rho_C$ values in the W dihydrogen complex are significantly lower than those in the Ir complex, in good agreement with the elongated nature of the Ir dihydrogen bond.

Further elongation of the H–H bond is observed in the binuclear dihydrogen ruthenium complex shown in Structure 3.2. In the absence of reliable structural data, the H–H distance in this complex has been determined from solution $^1\text{H}$ $T_{1\text{min}}$ measurements as 1.21 or 1.53 Å for a fast-spinning [around the Ru–(H₂)–Ru axis] or static (H₂) structural unit, respectively. However, the $^1\text{J(H–D)}$ constant of 15 Hz measured for the HD ligand corresponds to an H–H distance of 1.19 Å, closer to that of a fast-spinning unit. This example of dihydrogen bonding is particularly interesting because the Ru binuclear complex lies formally on the way to the intermolecular dihydrogen interactions M–H···H–Me that actually exist in solid transition metal hydride systems (see Section 7.3).

One of the important properties of dihydrogen ligands, particularly in charged transition metal complexes, is their ability to undergo heterolytic cleavage [9]. In addition, protonation of transition metal hydrides with acids is a common method for preparation of transition metal dihydrogen complexes:

$$M–H + HX \rightleftharpoons [M(H_2)]^+[X]^- \tag{3.3}$$

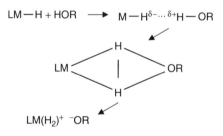

**Structure 3.2**   Structure proposed for the binuclear Ru dihydrogen complex in ref. 8, where the dihydrogen ligand is binding to two ruthenium atoms.

**Scheme 3.1**

Due to polarization of the M–H bond, the hydride ligand is usually negatively charged.

The process in eq. (3.3) is very close to proton transfer along the lines of that in classical hydrogen bonds. By analogy, it has been suggested that proton–hydride bonding interactions, $H^{\delta-}\cdots^{\delta+}H$, precede the protonation of transition metal hydrides through the formation of dihydrogen complexes, as shown schematically in Scheme 3.1 [10]. Here, the complex $H^{\delta-}\cdots^{\delta+}H$ is represented as an intermediate that is transformed to a dihydrogen complex via a corresponding transition state. Later, this proposition was supported by numerous experimental and theoretical studies in which the $H^{\delta-}\cdots^{\delta+}H$ interactions have been formulated as dihydrogen bonds.

It should be noted, however, that chemists had thought intuitively about proton–hydride interactions long before their formulation as dihydrogen bonds. For example, on the base of an x-ray crystal structure reported in 1934 for ammonium hypophosphite ($NH_4^+H_2PO_2^-$) it was suggested that hydrogen atoms in the hypophosphite group behave like $H^-$ ions with respect to ammonium [11]. However, this idea was refuted when restudy of the structure with accurate placement of hydrogen atoms demonstrated a lack of close H···H contacts [12]. This case illustrates problems that can appear in x-ray studies of proton–hydride interactions.

The first report in which $H^{\delta-}···^{\delta+}H$ interactions were truly recognized was published in 1970 by Brawn and co-workers [13]. On the basis of a thorough analysis of IR spectra recorded in $CCl_4$ solutions of boron coordination compounds $L–BH_3$ (where $L = Me_3N$, $Et_3N$, Py, or $Et_3P$) and $Me_3N–BH_2X$ (where $X = Cl$, Br, or I) in the presence of proton donors MeOH, PhOH, and p-$FC_6H_4OH$, the authors suggested a novel type of hydrogen bonding in which the $BH_3$ and $BH_2$ groups act as proton acceptors despite the absence of lone electron pairs or π-electrons. Figure 3.7 illustrates some of these IR spectra recorded in the ν(OH) regions. To avoid self-association processes, the spectra have been registered in low-concentration solutions. As is clear from the figure, the addition of aminoboranes reduces significantly the intensity of the free OH band. The latter

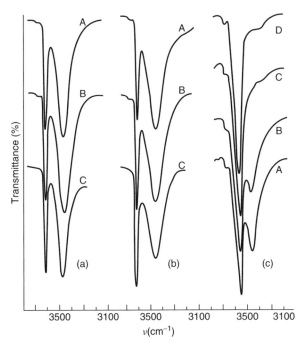

**Figure 3.7** The ν(OH) bands observed for (a) PhOH in the presence of $Me_3N–BH_3$ (A), $Et_3N–BH_3$ (B) $Et_3P–BH_3$ (C); (b) p-$FC_6H_4OH$ in the presence of $Me_3N–BH_2X$, where $X = Cl$ (A), $X = Br$ (B) $X = I$ (C); (c) p-$FC_6H_4OH$ in the presence of $Me_3N–BH_3$ (A), $Me_3N–BH_2Cl$ (B), $Me_3N–BHCl_2$ (C) in benzene and free p-$FC_6H_4OH$ (D). (Reproduced with permission from ref. 13.)

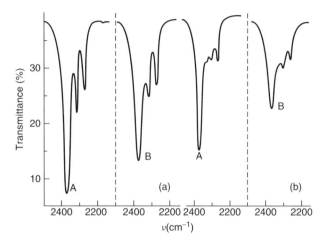

**Figure 3.8**   B–H absorptions observed for (a) $Me_3N–BH_3$ (0.1 M) in $CCl_4$ without (A) and with p-$FC_6H_4OH$ (0.2 M) (B); (b) $Me_2NH–BH_3$ in $CCl_4$ at concentrations of 0.002 mol/L (A) and 0.09 mol/L (B). (Reproduced with permission from ref. 13.)

is accompanied by the appearance of a broadened band at a lower frequency. This is typical of hydrogen bonding, as in the case of classical hydrogen bonds. A more complicated effect, caused by the presence of p-$FC_6H_4OH$, is observed for the $BH_3$ absorption of $Me_3N–BH_3$. The effect is shown in Figure 3.8. Here the strongest component in the IR spectra, assigned to antisymmetric stretching, undergoes the most change. The band intensity decreases with addition of the acidic component, whereas there is no observable change in the frequency of the peak maxima at 2273(m), 2319(m), and 2370(s) $cm^{-1}$. It should be added that perturbations of $BH_3$ absorption in an $Me_2N–BH_3$ molecule undergoing self-association in $CCl_4$ are very similar [see Figure 3.8(b)].

These first IR data show clearly that intermolecular interactions involve an electropositive hydrogen atom attached to an oxygen (or nitrogen) atom; but on the other hand, they indicate that the groups $BH_3$ and $BH_2$ participate in these interactions. However, questions about the nature of a proton-accepting center remain open. Despite this fact, variable-temperature IR studies have resulted in the first determinations of energy parameters characterizing dihydrogen bonding: The association energies have been found in the range 1.7 to 3.5 kcal/mol, comparable with those of conventional hydrogen bonds [14].

## 3.2. THE NATURE OF DIHYDROGEN BONDING: THE TOPOLOGY OF ELECTRON DENSITY AND CONTRIBUTIONS TO TOTAL BONDING ENERGY

Similarly to classical hydrogen bonding, the nature of dihydrogen bonds can be represented clearly in the framework of the AIM concept [15]. In the context of dihydrogen bonding, AIM theory deals with the same terms: charge density $\rho_C$, the Laplacian of the charge density, $\nabla^2\rho_C$, at bond critical points, and the topology of the bond paths. However, it now concerns the bond paths between the hydrogen atom in a proton donor and the hydrogen atom in a hydrogen-bond acceptor.

In three-dimensional space there are four types of critical points of rank 3, corresponding to nondegenerated points: termed $(3, -3)$, $(3, -1)$, $(3, +1)$, and $(3, +3)$ by theoreticians. The first and last types represent a maximum (which corresponds to a nuclear position) and a minimum, respectively. The two middle types represent saddle points called the *bond critical point* and the *ring critical point*, respectively. There is also a second set of special gradient paths conjugated to the bond paths. These gradient paths start at infinity but terminate at the bond critical point instead of being attracted to a nucleus. Therefore, this bundle of paths does not belong to any atom and forms a surface called the *interatomic surface* [16].

Since a topological analysis of electric density is available, first we summarize briefly various theoretical approaches and terms used in these approaches [2] (experimental methods are considered in Chapter 4).

1. *Molecular orbital an initio calculations*. These calculations represent a treatment of electron distribution and electron motion which implies that individual electrons are one-electron functions containing a product of spatial functions called *molecular orbitals:* $\Psi_1(x,y,z)$, $\Psi_2(x,y,z)$, and so on. In the simplest version of this theory, a single assignment of electrons to orbitals is made. In turn, the orbitals form a many-electron wave function, $\Psi$, which is the simplest molecular orbital approximation to solve Schrödinger's equation. In practice, the molecular orbitals, $\Psi_1$, $\Psi_2$,...are taken as a linear combination of $N$ known one-electron functions $\Phi_1(x,y,z)$, $\Phi_2(x,y,z)$:

$$\Psi_1 = \sum_{\mu=1}^{N} C_\mu \Phi_\mu \qquad (3.4)$$

These one-electron basis functions, $\Phi$, constitute the basis set. When the basis functions represent the atomic orbitals for the atoms in the molecule, eq. 3.4 corresponds to a linear combination of atomic orbitals (LCAO) approximation.

2. *Hartree−Fock (HF) calculations*. This approximation imposes two constraints in order to solve the Schrödinger equation and thus obtain the energy: (a) a limited basis set in the orbital expansion and (b) a single assignment of electrons to orbitals.

3. *Moller−Plesset (MP) perturbation approach*. This is an alternative treatment in the solution of the correlation problem. At a given basis set, the MP approach solves the full Hamiltonian matrix within Schrödinger's equation as the sum of two parts. Here, the second part represents a perturbation of the first

part. In the limits of the MP2 and MP4 calculations, the energy is expressed as a series, and correlation methods are formulated by truncation of the series to various orders. The truncation has been made after the second and fourth orders, corresponding to MP2 and MP4, respectively.

4. *Differential functional theory (DFT).* In the framework of the DFT approach, exact exchange for a single determinant in the HF method is replaced by a more general expression, known as the *exchange-correlation function*. This function can include terms accounting for both exchange energy and electron correlation, expressed, in turn, as a function of the density matrix. The latter is omitted from HF theory.

5. *B3LYP.* B3LYP is a hybrid method that includes a mixture of HF exchange and DFT exchange correlation. Functional B3 (Becke3) is a three-parameter exchange functional that contains the Slater exchange functional, HF and Becke's gradient correlation, and the LYP (Lee–Young–Parr) correlation functional.

6. *6-31G\* and 6-31G\*\*.* These are commonly used split-valence plus polarization basis sets. These basis sets contain inner-shell functions, written as a linear combination of six Gaussians, and two valence shells, represented by three and one Gaussian primitives, respectively (noted as 6-31G). When a set of six d-type Gaussian primitives is added to each heavy atom and a single set of Gaussian p-type functions to each hydrogen atom, this is noted as \* and \*\*.

7. *6-311++G\*\*.* This is a split-valence basis set plus polarization and diffuse functions. The basis comprises an inner shell of six S-type Gaussians and an outer (valence) region split into three parts that are represented by three, one, and one primitives, respectively. The latter is noted as 6-311G. The basis, supported by a single set of five Gaussians of the d-type for the first-row atoms and a single set of the p-type Gaussian functions for hydrogens is noted as \*\*. Incorporation of two sets of diffuse s- and p-functions corresponds to ++. Finally, it is worth mentioning that the MP2/6-311++G\*\* level of approximation seems to be one of the best compromises between calculation time and accuracy of results.

8. $E_{1BSSE}$. $E_{1BSSE}$ represents the interaction energy of hydrogen-bonded complexes which is corrected by using the basis set superposition error (BSSE). The interaction energies of all the hydrogen-bonded complexes are calculated as the difference between the total energy of the complex, AB, and the total energy of the isolated monomers, A and B: $E_1 = E_{AB} - (E_A + E_B)$. Since the interaction energies calculated are affected by the basis set superposition error, the latter can be estimated as

$$BSSE(A - B) = E(A)_A - E(A)_{AB} + E(B)_B - E(B)_{AB} \qquad (3.5)$$

Here $E(A)_{AB}$ is the energy of the monomer A obtained on the basis of its geometry within the dimer and the complete set of basis functions used to describe the dimer; and $E(A)_A$ is the energy of the same molecule but using only the basis functions centered on it. The BSSE problem, which is linked to the inconsistency of the basis sets and is used for the whole complex and for the monomers, may be explained in the following way. In the complex, electron distribution within each monomer is close to this distribution for the corresponding isolated species. However, in the complex, each monomer may compensate for the incompleteness of the basis set using the basis function of the neighbor. In such a situation the energy of the complex is lowered and the strength of the hydrogen bond is overestimated. Theoretically, BSSE is equal to zero in the complete basis set. When it is impossible to apply larger basis sets, it is necessary to use BSSE.

9. *Gauge-independent atomic orbitals.* GIAOs are eigenfunctions of a one-electron system that have been perturbed by an external magnetic field.

10. *Mulliken population analysis.* This treatment shows how a molecule under investigation distributes electrons according to the atomic orbital occupancy. Note that the overlap population between two atoms is divided evenly between them without consideration of possible differences in atom types, electronegativities, and so on.

Popelier [16] gives a very good example of the AIM description applied to the dimer structure of $BH_3-NH_3$. As has been shown, boron-coordinated compounds $L-BH_3$ are capable of hydrogen bonding in the presence of acidic components. However, a proton-accepting center has not been established experimentally. In this context, the $BH_3-NH_3$ molecule has a conventional proton donor, $N-H^{\delta+}$, and a potential proton acceptor, $B-H^{\delta-}$. The geometries of the monomer $BH_3NH_3$ and dimer $(BH_3-NH_3)_2$ structures have been optimized in the limits of the Hartree–Fock and MP2 approximations at the 6-31G** basis sets. The structure of dimer $(BH_3-NH_3)_2$ is depicted in Figure 3.9, and the geometric parameters selected are listed in Table 3.2.

Three relatively short contacts, $H(12)\cdots H(3)$, $H(5)\cdots H(15)$, and $H(16)\cdots H(6)$, are shown in Figure 3.9 with distances that are remarkably dependent on the computing method at the same basis set. For example, the $H(12)\cdots H(3)$ distance is calculated as 1.914 Å by the HF method versus 1.726 Å by the MP2 method. Despite this circumstance, all the $H\cdots H$ distances are smaller than the sum of the van der Waals radii of H (2.4 Å), thus suggesting dihydrogen bonding. The optimized geometrical parameters in Table 3.2 support this idea. As in the case of conventional hydrogen bonds, the data show elongations in the initial bonds B–H and N–H. However, the effects are unpronounced.

These effects are more significant for metal hydrides interacting with stronger proton donors (Table 3.3). Here elongations in the H–X bonds can reach 0.04 Å [17]. The data in Table 3.3 also show that the $\Delta r_{HX}$ values correlate with increasing the energies of dihydrogen bonds and decreasing the $H\cdots H$ distances from 1.378 Å to 1.692 Å.

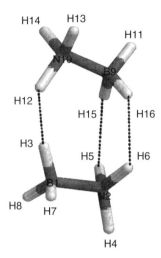

**Figure 3.9**  Dimeric structure $(BH_3-NH_3)_2$; the plane of the paper is the mirror plane. (Reproduced with permission from ref. 16.)

**TABLE 3.2. Selected Geometrical Parameters and Bonding Energies for the Dimeric Structure $(BH_3NH_3)_2$ Shown in Figure 3.9**

| Bond | Parameters [Å(deg)] at the Level: | |
| --- | --- | --- |
| | HF/6-31G** | MP2/6-31G** |
| N(2)–H(5) | 1.004 (1.003)[a] | 1.016 (1.014)[a] |
| N(10)–H(12) | 1.007 (1.003)[a] | 1.022 (1.014)[a] |
| B(9)–H(15) | 1.215 (1.208)[a] | 1.209 (1.202)[a] |
| B(1)–H(3) | 1.219 (1.208)[a] | 1.213 (1.202)[a] |
| H(3)···H(12) | 1.914 | 1.726 |
| H(6)···H(16) | 1.914 | 1.726 |
| H(5)···H(15) | 2.324 | 2.149 |
| N(10)–H(12)···H(3) | 163.5 | 168.7 |
| B(1)–H(3)···H(12) | 144.6 | 139.2 |
| B(9)–H(15)···H(5) | 113.8 | 112.6 |
| Energy (au) | −165.265 10 (−82.624 97)[a] | −165.885 50 (−82.932 29)[a] |

[a]Calculated for monomeric $BH_3NH_3$.

Going back to the dimer $(BH_3-NH_3)_2$, it should be emphasized that despite the remarkable influence of the computing method (note that the MP2 data illustrate the importance of the correlation effects) on H···H distances and N–H···H and B–H···H angles, the theoretical data correlate well with the purely geometric characteristics found experimentally by Richardson et al. in 18 structures, containing intermolecular dihydrogen contacts N–H···H–B [18].

**TABLE 3.3. H···H Bond Lengths, Elongations in Bond Lengths of Proton Donors HX, and Bonding Energies of Dihydrogen-Bonded Complexes**[a]

| Dihydrogen-Bonded Complex | $r$(H···H) (Å) | $\Delta r_{HX}$ (Å) | $E$ (kcal/mol) |
|---|---|---|---|
| Mo(NO)(CO)$_2$(PH$_3$)$_2$H···HF | 1.378 | 0.040 | −11.1 |
| Mo(NO)(CO)$_2$(NH$_3$)$_2$H···HOH | 1.647 | 0.013 | −13.1 |
| Cp*Ru(CO)(NO)H···HOH | 1.770 | 0.006 | −0.9 |
| CpRu(CO)(NO)H···HOCF$_3$ | 1.458 | 0.042 | −9.8 |
| LiH···HF | 1.611 | 0.021 | −10.93 |
| HMgH···HF | 1.582 | 0.012 | −6.02 |
| CpRu(CO)(PH$_3$)H···HOC(CF$_3$)$_2$ | 1.654 | 0.017 | −9.7 |
| CuH···HF | 1.692 | 0.013 | −6.8 |

[a]Obtained in the framework of B3PW91calculations.

The existence of short H···H distances is itself not enough to formulate these interactions as dihydrogen bonds. As we showed in Chapter 2, the more accurate criteria should be based on the topology of the electron density. Popelier has demonstrated that the dihydrogen bonds in (BH$_3$−NH$_3$)$_2$ correspond completely to hydrogen bond criteria: the topology of the electronic density, mutual penetration of the acidic hydrogen and the accepting hydrogen, increasing the net charge of the hydrogen atom, energetic destabilization of the hydrogen atom, decreasing dipolar polarization of the hydrogen atom, and decreasing hydrogen atomic volume. Some of these criteria can be seen clearly in Figures 3.10 and 3.11, showing the presence of bond critical points on all the H···H directions, which, in turn, correspond closely to the bond paths expected, associated with the H···H bond critical points.

Electron density $\rho_C$ calculated at the bond critical point and related to the *bond order* and *bond strength* is shown in Table 3.4. As shown, the $\rho_C$ values obtained are within the range 0.002 to 0.0035 au, magnitudes typical of conventional hydrogen bonds.

According to AIM principles, the Laplacian of the electron density, $\nabla^2{}_{\rho C}$, obtained at the bond critical points takes positive values for ionic bonds, hydrogen bonds, and van der Waals interactions called *closed-shell interactions*. By contrast, covalent bonds should exhibit large, negative $\nabla^2{}_{\rho C}$ values. In good agreement with AIM theory, the B−H and N−H bonds in the dimer (BH$_3$−NH$_3$)$_2$, for example, are clearly covalent, while the dihydrogen bonds found represent typical closed-shell interactions, with the Laplacian values lying in the expected range 0.014 to 0.139 au. The data in Table 3.4, obtained by two computing methods, support the reliability of the conclusions.

One aspect important for an understanding of the nature of dihydrogen-bonded complexes is their bonding energies, particularly different components that contribute to the total formation energies. The electrostatic, exchange, polarization, and charge transfer contributions have been calculated by Orlova and Schneider [19] for the four simplest hydrogen- and dihydrogen-bonded complexes:

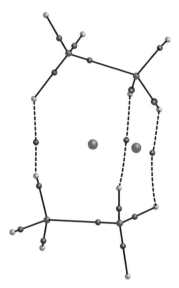

**Figure 3.10** Schematic representation of all the critical points in the dimeric structure $(BH_3-NH_3)_2$. The hydrogen atoms are represented by the small gray spheres, and the bond critical points are represented by the small black spheres. The ring critical points are noted as two large spheres. A dotted line shows the dihydrogen bond's bond path. (Reproduced with permission from ref. 16.)

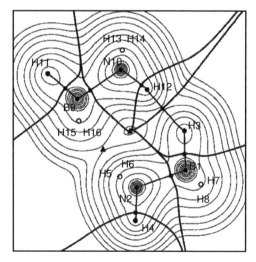

**Figure 3.11** Contour lines (gray) of the electron density, the molecular graph (black), and interatomic surfaces (black) in the dimeric structure $(BH_3-NH_3)_2$ obtained at the MP2/6-31G** level. Here the bond critical points are marked as squares and the ring critical points as triangles. The labels of the nuclei located in the mirror plane (the plane of the paper) are solid, and those that do not lie in this plane are open. (Reproduced with permission from ref. 16.)

**TABLE 3.4. Characterization of Bond Critical Points in the Dimeric Structure $(BH_3-NH_3)_2{}^a$**

| Level of Calculation | Bond | $\rho_C$ (au) | $\nabla^2 \rho_C$ (au) |
|---|---|---|---|
| HF | B(1)–H(3) | 0.1560 | −0.0007 |
| | N(10)–H(12) | 0.3557 | −2.0272 |
| | N(2)–H(5) | 0.3594 | −2.0395 |
| | B(9)–H(15) | 0.1614 | −0.0324 |
| | H(3)···H(12) | 0.0126 | 0.0353 |
| | H(15)···H(5) | 0.0067 | 0.0247 |
| MP2 | B(1)–H(3) | 0.1560 | 0.0221 |
| | N(10)–H(12) | 0.3429 | −1.9398 |
| | N(2)–H(5) | 0.3495 | −1.9720 |
| | B(9)–H(15) | 0.1622 | −0.0136 |
| | H(3)···H(12) | 0.0190 | 0.0464 |
| | H(15)···H(5) | 0.0096 | 0.0347 |

$^a$Obtained by HF and MP2 calculations.

HNO···HF, HNO···HOH, LiH···HF, and LiH···HOH. Table 3.5 summarizes the data obtained at the HF/6-31G* theoretical level. First, as in the case of conventional hydrogen bonds, the total energy of dihydrogen bonding depends on the strength of a proton donor and the total energy increases by 4 kcal/mol on going from Li–H···HOH to Li–H···HF. Second, the total energy gain obtained for dihydrogen bonds can be even higher than that for conventional hydrogen bonds. In fact, the energy increases by a factor of 2 from HNO···HF to Li–H···H–F. Third, electrostatic interactions provide the principal for both bonding types.

Nevertheless, the data in Table 3.5 reveal an important difference between classical hydrogen bonding and dihydrogen bonds. In fact, since in O···H conventional bonds the electrostatic component is followed by charge transfer energy while the polarization contribution $E_{PL}$ is very small, classical hydrogen bonds can be formulated as

$$E_{ES} > E_{CT} \qquad (3.6)$$

**TABLE 3.5. Energy Components (kcal/mol) for the Simplest Hydrogen- and Dihydrogen-Bonded Complexes$^a$**

| Complex | $E_{ES}$ | $E_{EX}$ | $E_{PL}$ | $E_{CT}$ | $E_{mix}$ | $E_{tot}$ |
|---|---|---|---|---|---|---|
| HNO···HF | −6.22 | 3.76 | −0.77 | −1.74 | 0.04 | −4.94 |
| LiH···HF | −13.24 | 7.95 | −2.75 | −3.17 | 0.69 | −10.52 |
| HNO···HOH | −3.06 | 1.69 | −0.26 | −1.02 | −0.02 | −2.66 |
| LiH···HOH | −10.97 | 7.88 | −1.13 | −2.42 | 0.36 | −6.27 |

$^a E_{ES}$, electrostatic; $E_{EX}$, exchange; $E_{PL}$, polarization; $E_{CT}$, charge transfer; $E_{mix}$; mixing. Calculated at the HF/6-31G* level.

**TABLE 3.6. Energy Components (kcal/mol) for Dihydrogen-Bonded Complexes[a]**

| Complex | $E_{ES}$ | $E_{EX}$ | $E_{PL}$ | $E_{CT}$ | $E_{mix}$ | $E_{tot}$ |
|---------|----------|----------|----------|----------|-----------|-----------|
| $(BH_3SH_2)_2$ | $-12.29$ | $11.36$ | $-4.95$ | $-3.59$ | $3.94$ | $-5.53$ |
| $(BH_3PH_3)_2$ | $-8.14$ | $6.73$ | $-3.54$ | $-1.52$ | $2.99$ | $-3.49$ |
| $(AlH_3PH_3)_2$ | $-9.08$ | $7.67$ | $-4.80$ | $-2.18$ | $4.45$ | $-3.94$ |

[a]Calculated at the MP2/6-31G* level.

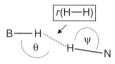

**Structure 3.3**

In contrast, dihydrogen bonds contain much larger $E_{PL}$ components, and thus can be shown as

$$E_{ES} > E_{CT} \approx E_{PL} \tag{3.7}$$

It should be noted that the $E_{PL}$ term reflects the distortion of electron distribution under complexation. In the case above, it shows changes in the initial Li–H and H–X bonds after the formation of H···H bonds. Finally, an analysis of the energy parameters in Table 3.6, calculated for dihydrogen bonding in dimer systems $(BH_3SH_2)_2$, $(BH_3PH_3)_2$, and $(BH_3PH_3)_2$ [20], shows that the polarization contributions, $E_{PL}$, can greatly surpass the charge transfer components, $E_{CT}$.

Bond lengths H···H and geometrical parameters of various dihydrogen-bonded complexes are discussed in subsequent chapters. Here, in the framework of general representations, it is reasonable to demonstrate a Crabtree analysis of crystallographic data taken from the Cambridge Structure Database [18]. Crabtree and co-workers found 18 x-ray structures of boron–nitrogen compounds containing 26 intermolecular contacts N–H···H–B with H···H distances smaller than 2.2 Å. The compounds found can be characterized by Structure 3.3, where the $\psi$ angles vary between 117° and 171°, with an average value of 149°. The $\theta$ angles have been found within the range 90 to 171°, with an average value of 120°. These parameters compare well with the data in Table 3.2. Moreover, such geometry is typical of dihydrogen bonds. Remarkable deviations from the geometry can be found only for *intramolecular* dihydrogen bonds, which are considered in Chapter 5.

## 3.3. SCALAR SPIN–SPIN COUPLING THROUGH DIHYDROGEN BONDS AS EVIDENCE OF THEIR PARTLY COVALENT CHARACTER

As pointed out in Chapter 2, electrostatic interactions strongly dominate in classical hydrogen bonds. Nevertheless, they still show a small covalent character, which can be deduced, for example, from scalar nuclear spin–spin coupling via hydrogen bonds observed in NMR spectra.

This $^1$H–$^1$H scalar spin–spin coupling through a dihydrogen bond has been investigated theoretically by Del Bane and co-workers [21] using models of linear and nonlinear dihydrogen-bonded complexes, such as NCH$\cdots$HLi, NHC$\cdots$HLi, NCH$\cdots$HNa, NHC$\cdots$HNa, LiNCH$^+\cdots$HLi, HOH$\cdots$HLi, and HOH$\cdots$HNa. The values of the $^1$J($^1$H–$^1$H) constants were strongly dependent on the nature of the proton donor/proton acceptor pairs and on their mutual orientation. The authors have evaluated all the terms that contribute to the total coupling constants.

According to NMR theory [30], contributions to spin–spin coupling come from paramagnetic spin–orbital (PSO), diamagnetic spin–orbital (DSO), Fermi contact (FC), and spin–dipolar (SD) interactions [30]:

$$H_{SO} = \frac{\mu_0}{4\pi} g_L \mu_B \gamma_N \hbar \sum 2 L_e I_N r_{eN}^{-3} \tag{3.8}$$

$$H_{SD} = \frac{\mu_0}{4\pi} g_S \mu_B \gamma_N \hbar \sum [r_{eN}^{-3} I_N S_e - 3 r_{eN}^{-5} (I_N r_{eN})(S_e r_{eN})] \tag{3.9}$$

$$H_{FC} = \frac{2}{3} \mu_0 g_S \mu_B \gamma_N \hbar \sum \delta(r_{eN}) I_N S_e \tag{3.10}$$

In these expressions, $e$ and $N$ refer to electron and nucleus, respectively, $L_e$ is the orbital angular moment operator, $r_{eN}$ is the distance between the electron and nucleus, $I_N$ and $S_e$ are the corresponding spins, and $\delta(r_{eN})$ is the Dirac delta function (equal to 1 at $r_{eN} = 0$ and 0 otherwise). The other constants are well known in NMR. It is worth mentioning that eqs. 3.8 and 3.9 show the interaction of nuclear spins with orbital and dipole electron moments. It is important that they not require the presence of electron density directly on the nuclei, in contrast to Fermi contact interaction, where it is necessary.

The total $^1$J($^1$H–$^1$H) values calculated for the linear CNH$\cdots$HLi complex from all the contributions above, are represented in Figure 3.12 as a function of the H$\cdots$H distance. Figure 3.13 illustrates the dependencies obtained for the nonlinear structure HOH$\cdots$H–Li. As shown, the SD terms are negligible at all H$\cdots$H distances, in good agreement with the data obtained for conventional hydrogen bonds. However, in contrast to the three-bond X–Y spin–spin coupling constants through conventional X–H–Y hydrogen bonds, which are determined completely by Fermi contact terms, the $^1$J($^1$H–$^1$H) couplings in the figures have nonnegligible contributions coming from both the PSO and DSO interactions. The calculations show that the total and Fermi contact terms change quadratically with the H$\cdots$H distance, while the PSO and DSO contributions correspond to linear dependencies. Finally, the PSO and DSO terms have similar magnitudes but opposite signs. For this reason, the two terms tend to cancel, and thus the shapes of the $^1$J($^1$H–$^1$H) curves are dictated completely by the Fermi contact terms.

As is clear from the figures, at small H$\cdots$H distances of $\approx 1$ Å, the total $^1$H–$^1$H constants are significant and determined to be 28 Hz for CNH$\cdots$HLi and 50 Hz for HOH$\cdots$HLi. It should be noted that these values are very close to

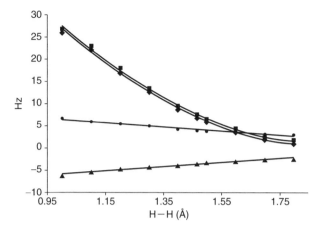

**Figure 3.12**  Total spin–spin coupling constant $^1J(H–H)$ (squares) and PSO (triangles), DSO (circles), and Fermi contact (diamonds) contributions as a function of the H$\cdots$H distance in the dihydrogen-bonded complex CNH$\cdots$HLi. (Reproduced with permission from ref. 21.)

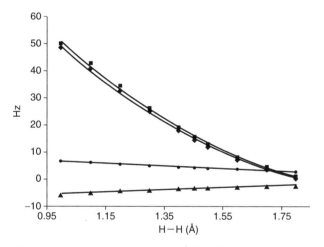

**Figure 3.13**  Total spin–spin coupling constant $^1J(H–H)$ (squares) and PSO (triangles), DSO (circles), and Fermi contact (diamonds) contributions as a function of the H$\cdots$H distance in the dihydrogen-bonded complex HOH$\cdots$HLi. (Reproduced with permission from ref. 21.)

those measured in elongated dihydrogen ligands of transition metal complexes [6]. However, the total $^1J(^1H–^1H)$ and Fermi contact terms then decrease with increasing H$\cdots$H separation. In addition, they change sign, occurring near the equilibrium H$\cdots$H distance. It is obvious that this effect can lead to a situation in which it would be difficult to measure the $^1H–^1H$ couplings experimentally in dihydrogen-bonded complexes because of their small predicted magnitudes. In good agreement with this statement, Shubina and co-workers [22] communicated that the hydride resonance of complex $WH(CO)_2(NO)(PMe_3)_2$ broadens in the presence of proton donor $(CF_3)_2CHOH$ by 2 to 4 Hz upon lowering the temperature and/or increasing the $(CF_3)_2CHOH$ concentration. The line width reaches a maximum of 7 Hz at $-100°C$, but no $^1J(^1H-^1H)$ constant is resolved. It should be emphasized that this behavior could be connected with equilibrium between free and coordinated hydride molecules, which is very fast on the NMR time scale. In fact, $^1H$ NMR experiments performed for the intramolecular dihydrogen-bonded complexes **1** shown in Structure 3.4 have revealed the $^1J(^1H^a-^1H^b)$ constants up to 5.5. Hz [23]. The $H\cdots H$ distances in **1** have been determined to be 1.7 to 1.8 Å. Let us consider this important result more attentively. Structure 3.5 illustrate schematically the structural evolution of the H–H unit from the free $H_2$ molecule to a classical metal dihydride with a large $H\cdots H$ distance via a dihydrogen complex with pronounced H–H bonding. In these terms it is obvious that the dihydrogen-bonded complex shown in Structure 3.6 would be located structurally between a dihydrogen complex with an elongated H–H bond and a classical dihydride.

It has been found experimentally [6] and then confirmed theoretically that between the H–H distance and the J(HD) spin–spin coupling in the partially

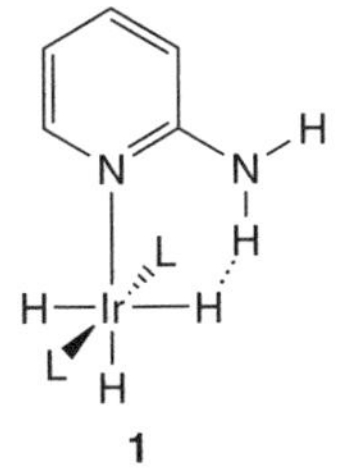

**1**

**Structure 3.4**

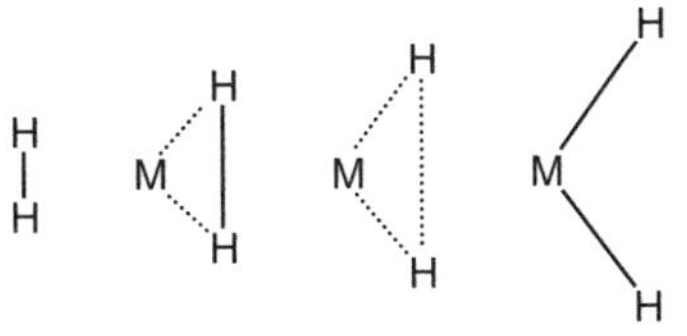

**Structure 3.5**

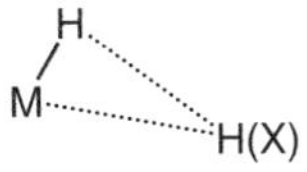

**Structure 3.6**

deuterated systems there is a good linear relationship:

$$r(\text{H–H})(\text{Å}) = 1.44 - 0.0168\text{J(H–D)} \qquad (3.11)$$

It is worth mentioning that eq. 3.11 has to be valid for $\text{J}(^1\text{H}-^1\text{H})$ to account for the $\gamma\text{H}/\gamma\text{D}$ ratio of 6.51. Finally, it has been established that the linearity of this expression breaks down when the H–H distances become longer than 1.3 Å. The bond-valence concept used for such systems and for their spin–spin coupling constants has resulted in the nonlinear but more common relationship given in ref. 24. It is remarkable that this relationship, showing correlation parameters similar to those obtained for strong hydrogen-bonded systems, predicts the J(H–H) constants to be 11 and 5.2 Hz for H$\cdots$H separation of 1.7 and 1.8 Å, respectively. The latter is in very good agreement with experiments performed for the iridium hydrides **1** in Structure 3.4. Thus, there is no doubt that dihydrogen bonding has a small covalent character, as has been established for classical hydrogen-bonded systems.

## 3.4. FIELD EFFECTS ON DIHYDROGEN BONDING

The nature of proton-accepting sites in classical hydrogen and nonconventional dihydrogen bonds is different. Nevertheless, the two types of bonding are similar. Since the formation energy for conventional hydrogen bonds depends on external pressure and polarity of environment, the same effects might be expected for dihydrogen bonds.

The electric field effects on conventional hydrogen-bonded complexes, particularly those on proton transfer along classical hydrogen bonds, have been analyzed by Ramos and co-workers [25]. The behavior of dihydrogen bonds in an external electric field has been traced by Rozas and co-workers [26]. It should be emphasized that the theoretical data on $H_2$ generation caused by the presence of an electric field may be important for an understanding of biological processes. In fact, similar reactions have been found for the enzyme hydrogenase, which catalyzes the activation of molecular hydrogen, leading to the uptake of $H_2$ gas.

Since A–H$\cdots$H–X systems can lose $H_2$, it is interesting to rationalize: Would the forces within a crystal be enough to generate molecular hydrogen from a dihydrogen bond? To provide an answer, the following neutral and charged systems,

$$\underset{\text{(a)}}{\text{Li–H}\cdots\text{H–F}} \rightleftharpoons \underset{\text{(b)}}{\text{Li}^+\cdots\text{H–H}\cdots\text{F}^-} \qquad (3.12)$$

$$H_3N\cdots H\text{–H}\cdots BH_3 \rightleftharpoons H_3N^+\text{–H}\cdots H\text{–}BH_3^- \qquad (3.13)$$

$$HBe\text{–H}\cdots H\text{–}NH_3^+ \rightleftharpoons HBeH^+\cdots H\text{–H}\cdots NH_3 \qquad (3.14)$$

have been calculated at the HF/6-31** level. The geometric parameters of the linear starting complexes have been optimized until stationary points were found.

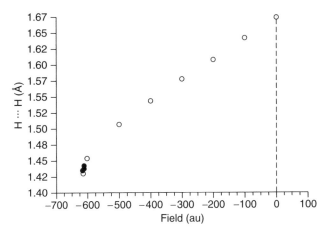

**Figure 3.14** Dihydrogen bond length (H···H) in system (3.12a) in the presence of an increasing electric field. (Reproduced with permission from ref. 26.)

$$
\begin{array}{c}
1.668 \text{ Å} \\
\text{Li}\!-\!-\!-\!\text{H}\cdots\text{H}\!-\!-\!-\!\text{F} \\
1.618 \text{ Å} \qquad 0.915 \text{ Å}
\end{array}
$$

**Structure 3.7**

For example, structure (a) of 3.12 (Figure 3.14), optimized in the absence of an electric field, represents a dihydrogen-bonded complex with a relatively short H···H distance (see Structure 3.7).

Due to the axial symmetry of these systems, a field can be applied along the $X$-axis in the positive and negative directions as well. Among the systems investigated, model (3.12), developed from a neutral dihydrogen-bonded molecule [state (a)] to a cation–$H_2$–anion hydrogen-bonded complex [state (b)] is most interesting. Since a *negative* field applied to system (a) can cause the formation of the $H_2$ molecule, this system was placed into an electric field that varied from $-0.00557$ to $-0.03342$ au (1 au of field is equal to $5.142 \times 10^{11}$ V/m). Figure 3.14 illustrates the pattern of H···H distance versus electric field, where the distance decreases in an asymptotic curve and reaches a limit of 1.4 Å. In accord, the Li–H and H–F bond lengths increase from 1.644 Å to 2.072 Å and from 0.920 Å to 0.976 Å, respectively. The latter leads to elongation in the Li···F distance to 4.5 Å (Figure 3.15). When the external electric field becomes larger than $-0.03342$ au, both heavy atoms tend to separate infinitely from the $H_2$ molecule in the middle. It is obvious that in the case of charged system (b) in eq. 3.12, an electric field can be applied in both directions, leading to different results: breaking of the H–H bond or approximation of both ions to the $H_2$ molecule. It has been found that the $H_2$ molecule in system (b), placed in a *negative* field, transforms immediately to a dihydrogen bond with an H···H distance of $\approx 1.6$ Å. Further increase in the electric filed reduces the H···H distance and at a field of $-0.0228$

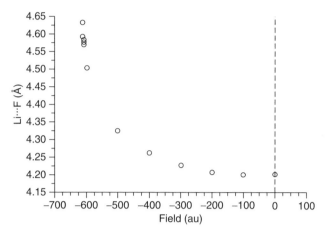

**Figure 3.15**  Effects of the external electric field on the Li· · ·F distance in system (3.12a). (Reproduced with permission from ref. 26.)

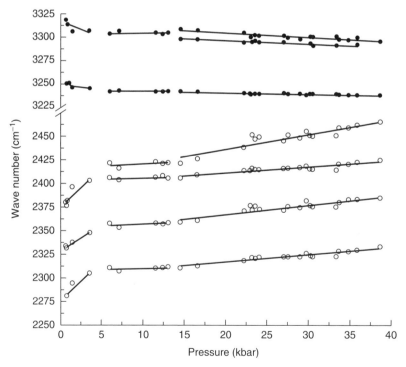

**Figure 3.16**  Effects of pressure on the N–H and B–H stretching frequencies in the Raman spectra of an $NH_3BH_3$ molecule. (Reproduced with permission from ref. 28.)

au the dihydrogen bond becomes covalent and both $Li^+$ and $F^-$ ions segregate. In a *positive* field, system (b) becomes similar to system (a), where the H···H distance increases up to a positive field of 0.03342 au, when LiH and HF tend to separate to the infinite.

The strong effects of an external electric field on dihydrogen bonding and proton transfer found theoretically provide the similar processes expected within the crystal due to internal forces. On the other hand, they are related to the solvent effects in solution. For example, according to high-level DFT calculations, a dihydrogen bonding energy between $HOC(CF_3)_3$ and the hydride ligand of $(Cp)Ru(CO)(PH_3)H$ determined in the gas phase as $-9.7$ kcal/mol decreases to $-5.6$ kcal/mol in *n*-heptane and to $-2.4$ kcal/mol in $CH_2Cl_2$ [27]. A similar tendency observed experimentally has been reported by Shubina and co-workers [22].

## 3.5. PRESSURE EFFECTS ON DIHYDROGEN BONDING

Upon the formation of a classical hydrogen bond, the hydrogen stretching frequencies of proton donors, O–H or N–H, are known to undergo relatively large shifts to lower wave numbers. Commonly, the vibrational frequencies in hydrogen-bonded systems X–H···Y are strongly dependent on the X···Y separations. In turn, these separations can be reduced by the application of high external pressures, which would be observed in the vibrational spectra. In reality, the pressure effects observed for classical hydrogen bonds are not simple. Generally, the frequencies decrease until the hydrogen bonds are symmetrical, and then they increase. The effects of pressure on dihydrogen bonding have been reported by Trudel and Gilson [28], who have studied the Raman spectra of the $BH_3NH_3$ molecule under pressure up to 40 kbar to confirm that dihydrogen bonds are actually formed.

The dependencies on the pressure of the bands corresponding to the N–H and B–H stretching modes are shown in Figure 3.16. As can be seen, the behavior of the N–H and B–H bonds is different: The slopes of the N–H antisymmetric and symmetric stretching frequencies are negative, whereas they are positive for the corresponding B–H stretches. According to x-ray data, structures of amine boranes with intermolecular dihydrogen bonds (see Structure 3.3) show N–H···H and B–H···H angles in the range 150 to 170° and 95 to 115°, respectively. In other words, the N–H···H fragments are more linear and the B–H···H angles are more acute. In this connection, the authors believe that the extension of the N–H bond is directed not at the (B)–H atom but at the electron-dense region of the σ -bond B–H. Then the pressure-induced decrease in distance leads to increasing the strength of the dihydrogen bonds at the expense of the N–H bond. On the other hand, since the (B)–H atom is not directly involved in the dihydrogen bonding, the slopes observed for the B–H stretching frequencies are positive.

An alternative explanation of negative pressure dependencies can be found in the work of Li and co-workers [29], who have suggested that the dihydrogen bond involves donation of charge from the σ-bond B–H to the N–H σ*-orbital.

Then an increase in antibonding character could reduce the bond strength and the force constant of stretching vibration. It should be emphasized that despite the variety of interpretations, pressure effects are observed clearly and confirm independently the existence of dihydrogen bonds.

## 3.6. DIFFERENCE BETWEEN HYDROGEN AND DIHYDROGEN BONDS

As mentioned above, topological analysis of the electronic density at critical points on a bond is the most useful instrument for the characterization of hydrogen and dihydrogen bonds. Despite the similarity of these two types of bonding, this approach allows us to determine the important difference between them. To show this difference, Hugas and co-workers have recently performed B3LYP/6-31++G(d,p) and MP2/6-31++G(d,p) calculations for various classical hydrogen- and dihydrogen-bonded complexes [31]. The authors have probed hydrogen bonds from very weak (HCl$\cdots$HBr) to strong (NH$_3$$\cdots$HF), with B3LYP

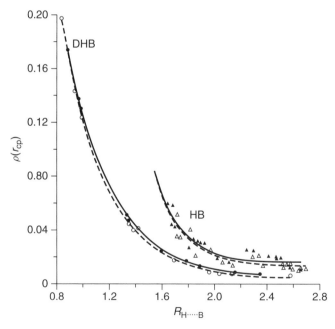

**Figure 3.17** Electron density at the bond critical point (in au) versus intermolecular distance (in Å) obtained for hydrogen-bonded (HB) and dihydrogen-bonded (DHB) complexes. Solid circles and triangles correspond to B3LYP calculations. Open circles and triangles represent MP2 calculations. The solid and dashed lines are fittings to the exponential function for B3LYP and MP2, respectively. (Reproduced with permission from ref. 31.)

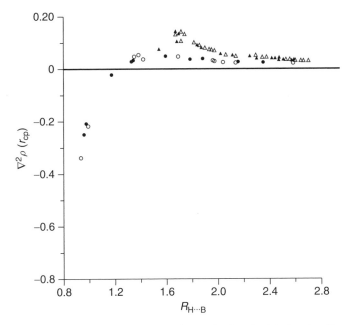

**Figure 3.18** Laplacian of the electron density at the bond critical point (in au) versus intermolecular distance (in Å). Solid circles and triangles represent B3LYP calculations performed for dihydrogen- and hydrogen-bonded complexes, respectively. Open circles and triangles correspond to MP2 calculations. (Reproduced with permission from ref. 31.)

bonding energies of $-1.06$ and $14.02$ kcal/mol, respectively. Dihydrogen-bonded complexes have also been represented by various systems from weak ($H_2BH\cdots$ HCl; $\Delta E = -0.6$ kcal/mol) to strong (NaH$\cdots$HBr; $\Delta E = -17.78$ kcal/mol). First, on the structural level it has been established that H$\cdots$H interactions exhibit shorter dihydrogen bond lengths than those of hydrogen-bonded complexes that have the same strength. Second, the topological properties for hydrogen and dihydrogen bonding where dependence on the theory level was insignificant have also been found to differ. Figure 3.17 shows the relationships between the $\rho_C$ values and the intermolecular distances, which appear to be *different* for the set of dihydrogen-bonded complexes and the set of hydrogen-bonded systems. However, these dependencies can be well fitted to the exponent functions for both types. It follows from the patterns that dihydrogen-bonded complexes are characterized by smaller electron density parameters than those for classical hydrogen bonds. The Laplacian values are illustrated in Figure 3.18. Again, the hydrogen bonds show higher $\nabla^2\rho_C$ values than those of dihydrogen-bonded complexes. The more interesting conclusion can be made from the sign of the Laplacian magnitudes. All the hydrogen bonds investigated exhibited a positive Laplacian, although strong dihydrogen bonds with bonding energies close to those of hydrogen-bonded complexes can show a negative Laplacian.

## 3.7. CONCLUDING REMARKS

1. Dihydrogen bonds, $H^{\delta-}\cdots^{\delta+}H$, are formed by two hydrogen atoms that have opposite charges when the distance between them is smaller than the sum of the van der Waals radii of H (2.4 Å).

2. Interactions between two hydrogen atoms separated by a distance of less than 2.4 Å can be formulated as dihydrogen bonds if they correspond to criteria based on the topology of the electron density in the $H\cdots H$ directions: The $\rho_C$ values should be small, and the $\nabla^2{}_{\rho C}$ values should be small and positive.

3. Bonding energies are similar for classical hydrogen bonds and dihydrogen bonds.

4. As in the case of classical hydrogen bonds, electrostatic contributions dominate in energies of dihydrogen bonding. However, in contrast to classical hydrogen bonds, where the electrostatic component is followed by charge transfer energy, $E_{CT}$, with a very small polarization contribution, $E_{PL}$ the dihydrogen bonds contain much larger $E_{PL}$ components, and thus $E_{ES} > E_{CT} \approx E_{PL}$.

5. $Y-H^{\delta-}\cdots^{\delta+}H-X$ bonds generally show linear $H\cdots H-X$ fragments (or close to linearity), whereas the $Y-H\cdots H$ fragments are bent.

6. The pressure and external electric field affect dihydrogen bonds.

7. Dihydrogen bonds show a small covalent character.

8. What makes this bond so unusual is that the proton-accepting atom is a hydrogen atom, although the topological properties of the electron density at the bond's critical points allow us to distinguish these two types of bonding.

## REFERENCES

1. J. Wessel, J. C. Lee, E. Peris, G. P. A. Yap, J. B. Fortin, J. S. Ricci, G. Sini, A. Albinati, T. F. Koetzle, O. Eisenstein, A. L. Rheingold, R. H. Crabtree, *Angew. Chem. Int. Ed. Engl.* (1995), **34**, 2507.

2. I. Alkorta, I. Rozas, J. Elguero, *Chem. Soc. Rev.* (1998), **27**, 163.

3. D. Braga, F. Grepioni, *Acc. Chem. Res.* (2000), **33**, 601.

4. K. M. Robertson, O. Knop, T. S. Cameron, *Can. J. Chem.* (2003), **81**, 727.

5. G. J. Kubas, *Acc. Chem. Res.* (1988), **21**, 120.

6. D. M. Heinekey, A. Lledos, J. M. Lluch, *Chem. Soc. Rev.* (2004), **33**, 175.

7. F. Maseras, A. Lledos, M. Costa, J. M. Poblet, *Organometallics* (1996), **15**, 2947.

8. J. P. Collman, P. S. Wagenknecht, J. E. Hutchison, N. S. Lewis, M. A. Lopez, R. Guilard, M. L'Her, A. A. Bothner-By, P. K. Mishra, *J. Am. Chem. Soc.* (1992), **114**, 5654.

9. R. H. Morris, in *Recent Advances in Hydride Chemistry*, ed. M. Peruzzini and R. Poli, Elsevier, New York (2001), pp.1–38.

10. S. Feracin, T. Burgi, V. I. Bakhmutov, E. V. Vorontsov, I. Eremenko, A. B. Vimetits, H. Berke, *Organometallics* (1994), **13**, 4194.

11. W. H. Zachariasen, R. C. L. Mooney, *J. Chem Phys.* (1934), **2**, 34.

12. R. Custelcean, J. E. Jackson, *Chem. Rev.* (2001), **101**, 1963.

13. M. P. Brown, R. W. Heseltine, P. A. Smith, P. J. Walker, *J. Chem. Soc. A* (1970), 410.

14. M. P. Brown, P. J. Walker, *Spectrochim. Acta* (1974), **30A**, 1125.

15. R. F. W. Bader, *Atoms in Molecules: A Quantum Theory*, Oxford University Press, New York (1990).

16. P. L. A. Popelier, *J. Phys. Chem. A* (1998), **102**, 1873.

17. N. V. Belkova, E. S. Shubina, L. M. Epstein, *Acc. Chem. Res.* (2005), **38**, 624.

18. T. B. Richardson, S. De Gala, R. H. Crabtree, P. E. M. Siegbahn, *J. Am. Chem. Soc.* (1995), **117**, 12875.

19. G. Orlova, S. Schneider, *J. Phys. Chem. A* (1998), **102**, 260.

20. S. A. Kulkarni, A. K. Srivastava, *J. Phys. Chem. A* (1999), **103**, 2836.

21. J. E. Del Bene, S. A. Perera, R. J. Bartlett, I. Alkorta, J. Elguero, O. Mo, M. Yanez, *J. Phys. Chem. A* (2002), **106**, 9331.

22. E. S. Shubina, N. V. Belkova, A. N. Krylov, E. V. Vorontsov, L. M. Epstein, D. G. Gusev, M. Niedermann, H. Berke, *J. Am. Chem. Soc.* (1996), **118**, 1105.

23. E. Perris, J. C. Lee, J. R. Rambo, O. Eisenstein, R. H. Crabtree, *J. Am. Chem. Soc.* (1995), **117**, 3485.

24. S. Grudermann, H. H. Limbach, G. Buntkowsky, S. Sabo-Etienne, B. Chaudret, *J. Phys. Chem. A* (1999), **103**, 4752.

25. M. Ramos, I. Alkorta, J. Elguero, N. S. Golubev, G. S. Denisov, H. Benedict, H. H. Limbach, *J. Phys. Chem. A* (1997), **101**, 9791.

26. I. Rozas, I. Alkorta, J. Elguero, *Chem. Phys. Lett.* (1997), **275**, 423.

27. M. Besora, A. Lledos, F. Maseras, N. V. Belkova, L. M. Epstein, E. S. Shubina, FIGIPS meeting, Barcelona, Spain (2001).

28. S. Trudel, D. F. R. Gilson, *Inorg. Chem.* (2003), **42**, 2814.

29. J. Li, F. Zhao, F. Jing, *J. Chem. Phys.* (2002), **116**, 25.

30. R. K. Harris, *Nuclear Magnetic Resonance Spectroscopy A Physicochmical View*, Longman Scientific & Technical, Harlow, England (1986), pp. 218–220.

31. D. Hugas, S. Simon, M. Duran, *J. Phys. Chem. A.*

# 4

# HOW TO FIND A DIHYDROGEN BOND: EXPERIMENTAL CRITERIA OF DIHYDROGEN BOND FORMATION

The theoretical criteria for the formation of dihydrogen-bonded complexes are based on the geometry, energy, and the topology of electronic density on H···H directions; charge density, $\rho_C$; and the Laplacian of the charge density, $\nabla^2 \rho_C$, at the bond critical points, and do not depend on crystallographic, spectroscopic, or other experimental data. However, it is obvious that despite the very high level of accuracy obtained using modern computing methods, theoretical results require independent experimental confirmation. As we show below, even the electron density parameters $\rho_C$ and $\nabla^2 \rho_C$ can be investigated experimentally by a very precise x-ray technique. Moreover, theoreticians often simplify objects under investigation to avoid numerous problems appearing in the calculation of many-electron systems. The validity of such simplifications also requires experimental support.

One important problem in theoretical studies is the remarkable dependence of calculated geometrical and energy parameters on the methods and/or basis sets that have been used. This is particularly important for H···H distances, which change significantly on going from one theory level to another. Finally, theoreticians have to be sure that an optimized structure truly occupies an energy minimum on the potential energy surface. It is well known that frequency analysis helps to solve this problem. For example, the presence of even one negative (imaginary) frequency can reveal the nature of a molecular system as a transition state. However, even in this case, one method can describe the system as a structure located at an energy minimum, whereas a second method can formulate the same system as a transition state. We consider such examples in subsequent chapters.

The majority of theoretical criteria enumerated in Chapter 3, can be reformulated in an experimental context as follows. Dihydrogen-bonded complexes are formed if:

1. An H···H distance is less than a sum of the van der Waals radii of H (2.4 Å).

*Dihydrogen Bonds: Principles, Experiments, and Applications*, By Vladimir I. Bakhmutov
Copyright © 2008 John Wiley & Sons, Inc.

2. An H–H interaction has directionality.

3. Initial bonds in the proton donor and proton acceptor sites elongate on dihydrogen bonding.

Points 1 and 2, which relate to the geometry of complexes, can be seen directly from x-ray or neutron diffraction structures. Distances H· · ·H in solution can be determined by proton spin–lattice NMR relaxation experiments. Point 3, which relates to changes in proton-donor and proton-acceptor molecules, can be probed by vibrational techniques employed for solutions, in the solid state and the gas phase. NMR spectra reflecting perturbations in magnetic shielding at target nuclei can be used for studies of dihydrogen bonding in solution. It is obvious that the choice of experimental method depends significantly on the nature of dihydrogen-bonded complexes and their aggregate state. However, a combination of several experimental approaches will probably provide the most reliable conclusions.

Commonly, any experimental study of dihydrogen bonds is undertaken to clarify the following aspects: (1) establishment of intra- or intermolecular coordination, (2) determination of its stoichiometry, (3) reliable establishment of a proton-donor site and a proton-acceptor center, (4) description of the geometry of dihydrogen-bonded complexes, and (5) correct measurement of bonding energies. In this chapter we demonstrate how to approach these factors using various experimental methods that work in the solid state, the gas phase, and in solution.

## 4.1. DIHYDROGEN-BONDED COMPLEXES IN THE SOLID STATE: X-RAY AND NEUTRON DIFFRACTION EVIDENCE

Since x-ray crystallography provides a "visualization" of a molecular image, it is difficult to overestimate its role in structural chemistry. For this reason, x-ray diffraction continues to be the method of choice for structural investigations of molecules and molecular assemblies containing hydrogen or dihydrogen bonds.

It is obvious that these molecular systems can be prepared specially or can form spontaneously as self-assemblies when hydrogen or dihydrogen bonds dictate the processes of molecular aggregation. The latter is particularly important in supramolecular chemistry and crystal engineering, where the primary attention is focused on mastering weak intermolecular interactions [1].

First, let us comment briefly on how to distinguish van der Waals interactions and hydrogen or dihydrogen bonding. This is not a simple question because there is no restrictive border between weak bonding and van der Waals interaction. The fundamental difference between these intermolecular interactions is not their energy but their different *directionality* or its absence. As we have shown above, hydrogen and dihydrogen bonding are inherently directional, showing a linear (or close to linear) geometry. The latter is favored energetically over bent geometries. In contrast, the van der Waals interactions are always *isotropic*. Here, interaction energies are independent of the contact angles. However, in practice, application of this principle is notsimple because hydrogen and dihydrogen bonds are directionally soft. For this reason, even relatively strong hydrogen or dihydrogen bonds cannot be characterized by a single example. A more accurate description of angular preferences requires statistical analysis of large quantities of structural data that can be extracted, for example, from the Cambridge Structural Database [2].

In addition to the geometrical softness of hydrogen- and dihydrogen-bonded complexes, localizations of small hydrogen atoms represent a well-known problem in x-ray experiments. This can be illustrated by the x-ray diffraction study of $NH_4H_2PO_2$ already mentioned, where the existence of proton–hydride interactions was first suggested, in 1934 [3]. In fact, the molecule $NH_4H_2PO_2$ has actually been found in the solid state as a self-associate. Phosphorus in this molecule is less electronegative than hydrogen, thus having a hydridic character. In contrast, the N–H bond in the ammonium ion could be a proton donor. However, because of the absence of reliable localizations of hydrogen atoms in the structure, this logical conclusion was wrong. Marincean and co-workers [4] have reinvestigated this molecule by the modern x-ray technique and accurately localized the hydrogen atoms. The structure is shown in Figure 4.1, where the NH$\cdots$HP distances are determined as 2.79 and 2.82 Å. Since they are significantly larger than the sum of van der Waals radii of H, dihydrogen bonds can be ruled out. At the same time, it is seen that the solid-state network is formed by the classical hydrogen bonds N–H$\cdots$O, in good agreement with the molecular orbital calculations performed.

Accurate localization of hydrogen atoms is particularly important when small hydrogen atoms are located near heavy atoms: for example, transition metals. Generally, it is now well established that x-ray crystallography usually underestimates H$\cdots$H and Me$\cdots$H distances by 0.2 to 0.3 Å [5]. In this sense, even $^1$H NMR relaxation experiments performed in solution can lead to more plausible

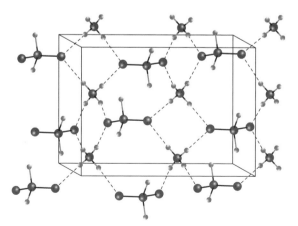

**Figure 4.1**   X-ray structure of $NH_4H_2PO_2$. Conventional N–H$\cdots$O bonds are shown as dashed lines. (Reproduced with permission from ref. 4.)

results that we get with neutron diffraction data [6]. The latter is particularly useful for characterizations of dihydrogen bonding. Finally, some atoms if similar size can be entangled in the x-ray experiments, thus leading to confusing structures. In such cases, neutron diffraction experiments are particularly useful.

Despite these intrinsic technical problems, a crystallographic structure of good quality speaks for itself. Therefore, in this chapter we discuss some representative structures that on the one hand, characterize different classes of dihydrogen-bonded complexes, and on the other, provide reliable general conclusions.

Dihydrogen bonds play a particularly important role in the chemistry of transition metal hydride complexes. These molecules provide a negatively polarized hydride ligand that acts as a strong proton acceptor, forming a dihydrogen-bonded complex which then participates in a proton transfer reaction to give a $(H_2)-M$ system. However, a priori, it was unclear what center accepts the proton: the metal–hydrogen $\sigma$-bond or the nonbonding $d_\pi$-electrons in the metal atoms [7]. This fundamental question can be solved by crystallographic data collected for systems where a proton-acceptor center has no *nonbonding valence electrons*. As mentioned earlier, different amine boranes, particularly $BH_3NH_3$, were key candidates in the studies performed by Richardson and co-workers [8]. Eighteen x-ray structures actually showned short (<2.2 Å) intermolecular contacts $N-H\cdots H-B$. To minimize the systematic errors usually associated with hydrogen positions determined by the x-ray method, the authors have normalized the $N-H$ and $B-H$ bond lengths as 1.03 and 1.21 Å, respectively. The majority of the structures treated (Figure 4.2) reveal the *directionality* of the $H\cdots H$ interactions

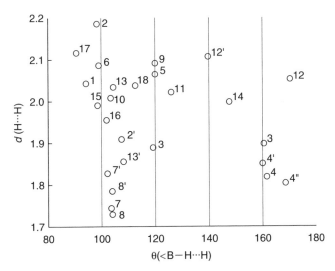

**Figure 4.2**   Proton–hydride $(H\cdots H)$ distances (Å) versus $B-H\cdots H$ angles (degrees) in the Cambridge Structure Database for 18 x-ray structures of boron nitrogen compounds. The numbers correspond to the CSD file names, which can be found in ref. 7. (Reproduced with permission from ref. 8.)

with N–H···HB angles between 117 and 171°, whereas the NH···H–B angles were more bent (95 to 120°). Thus, H···H distances below 2.2 Å and the geometrical features noted above can be taken as x-ray criteria for the formation of dihydrogen bonds.

It is interesting that x-ray data collected for the $BH_3NH_3$ molecule have confirmed the presence of bonding N–H···H–B. However, the geometry of the bonding has shown opposite tendencies: The B–H···H and N–H···H angles were linear and bent, respectively. It became obvious that the earlier assignment of boron and nitrogen atoms is reversed. This idea has been supported completely by neutron diffraction studies of $BH_3NH_3$ (Figure 4.3).

As already mentioned, x-ray diffraction does not reflect the X–H bond lengths properly because the positions of atoms are determined from the *electron density maxima*. Hence, localization errors are connected primarily with the positions of hydrogen atoms. Even the C–H bond lengths found in x-ray structures should be corrected to 1.083 Å. In other words, hydrogen atoms being compared, for example, to oxygen or carbon atoms are weak scatterers of x-rays. The situation changes in the case of neutron scattering. For this reason, neutron diffraction is the best method for accurate determinations of hydrogen positions in the solid state. According to neutron diffraction data, the compound $BH_3NH_3$ actually shows three short H···H contacts: $H_1···H_4$, $H_2···H_3$, and $H_2···H_4$, determined as 2.21(4), 2.02(3), and 2.23(4) Å, respectively, the smallest being $H_2···H_3$. In addition, the $H_2···H_3$–B and N–$H_2···H_3$ angles are measured as 106(1) and 156(3)°. It is obvious that this is best method for visualization of dihydrogen bonds existing in the solid state. The method has been employed to provide absolute

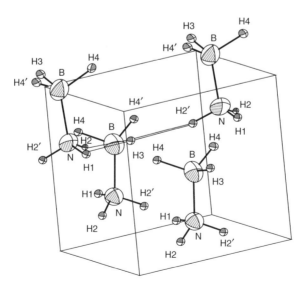

**Figure 4.3**   Neutron diffraction structure of $BH_3NH_3$. (Reproduced with permission from ref. 8.)

**Structure 4.1** Schematic presentation of the neutron diffraction structure, showing the solid-state self-assembly of cyclotrigallazane. (From ref. 9.)

structural evidence for the existence of dihydrogen bonds in self-assemblies, shown in Gladfelter's structure in Structure 4.1. Here, intermolecular distances N–H···H–Ga have been measured as 1.97 Å, again smaller than the van der Waals sum. It is interesting that the NH···H–Ga and N–H···HGa angles in this structure were similar (131 and 145°, respectively).

Remaining in the framework of this group, it is pertinent to show the x-ray structure of the dihydrogen-bonded complex where the polyhedral *closo*-$[B_{10}H_{10}]^{2-}$ anion accepts a proton of $CH_3OH$ [10]. As shown in Figure 4.4, the geometry observed is again typical of dihydrogen bonding; the O–H···H–B distance is measured as 2.2(2) Å and the O(1 S)H(1 S)···H(10) and (H1 S)···H (10)–B angles are equal to 171(18) and 116(12)°, respectively.

Among the x-ray and neutron diffraction characterizations of intermolecular dihydrogen bonding in transition metal hydrides, the indole···$ReH_5(PPh_3)_3$ complex is probably most representative. The complex has been prepared by cocrystallization with indole [11]. It is worth mentioning that a control cocrystallization of the Re complex with indene in which the acidic NH proton is replaced by the $CH_2$ group did not lead to a coordination product. The fragment of the neutron diffraction structure in Figure 4.5 shows clearly the presence of a dihydrogen Re–H···H–N interaction which could be represented as a three-center hydrogen bond. At the same time, one hydrogen ligand is bonded more strongly, with an H···H distance of 1.734 Å. The second ligand is remote by 2.212 Å. As in the case of $BH_3NH_3$, the Re–H···HN angles are remarkably bent (118.9 and 97.2° for the stronger and weaker bonds, respectively).

Unpronounced polarization of regular C–H bonds such as methyl or ethyl groups does not lead us to expect the formation of dihydrogen bonds. Actually, an analysis of the Cambridge Structure Database [2] reveals numerous C–H···H–C contacts with H···H distances of 2.500 Å or even more where the angular distribution is ideally isotropic. The latter is typical of nondirectional van der Waals interactions. The C–H and B–H bonds in the molecule 1,3-bis(2,4,6-trimethylphenyl)imidazol-2-ylidene· $BH_3$ are obviously more

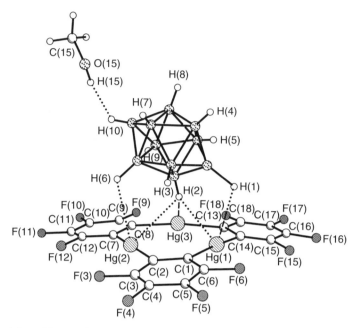

**Figure 4.4**  X-ray single-crystal structure of the cyclic trimeric perfluoro-*o*-phenylenemercury, coordinating the polyhedral *closo*-$[B_{10}H_{10}]^{2-}$ anion, showing dihydrogen bonding with $CH_3OH$. The structure was solved by the direct method and refined by the full-matrix least-squares technique against $F^2$ using anisotropic temperature factors for all nonhydrogen atoms. All the hydrogen atoms of the borate anion as well as the hydroxyl hydrogen of the methanol solvate molecule were located using Fourier synthesis and refined by isotropic approximation. (Reproduced with permission from ref. 10.)

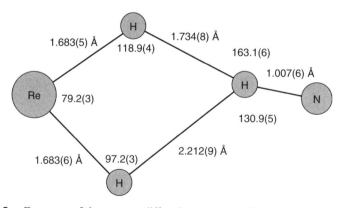

**Figure 4.5**  Fragment of the neutron diffraction structure of the adduct $ReH_5(PPh_3)_3\cdots$ indole. (Reproduced with permission from ref. 11.)

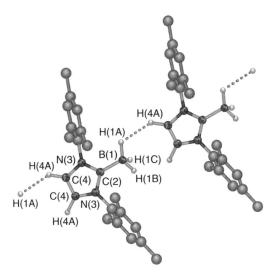

**Figure 4.6** X-ray structure of 1,3-bis(2,4,6-trimethylphenyl)imidazol-2-ylidene·BH$_3$. (Reproduced with permission from ref. 12.)

polarized. As result, the x-ray structure of this adduct, shown in Figure 4.6, reveals the C–H···H–B contact of 2.24 Å (see the dotted lines). In addition, these contacts have a pronounced directionality: The B–H···HC and BH···H–C angles are determined to be 113 and 138.5°, respectively [12].

### 4.1.1. Topology of Electron Density in Dihydrogen-Bonded Systems from Diffraction Data

Initially, the topological approach has been widely used only for the analysis of *theoretical* charge densities. Theoretical analysis allows us to consider two atoms as bonded, whereas experiments show only their spatial proximity (x-ray or neutron diffraction), changes in the vibrational behavior of an H–X bond involved in interaction, changes in magnetic shielding of target nuclei (NMR), and so on.

The first x-ray experiments with charge density were carried out in the late 1960s and 1970s to show the accumulation of electron density in bonding and nonbonding regions of molecular crystals. Later development of multipole models, more accurate data collection, faster computers, and modernization of instrumentation through the widespread use of area detectors revolutionized the charge density field by reducing data collection times from weeks to days or even hours [13]. The topological approach can now be applied to x-ray diffraction charge densities due to various multipole techniques that provide analytical evaluation of the Laplacian, $\nabla^2 \rho_C$, and gradient vector field of $\rho$, $\nabla \rho$. Despite the strong development of this field, use of this technique for studies of dihydrogen bonding is only beginning. Here we discuss the investigation of monohydride *cis*-HMn(CO)$_4$PPh$_3$ by Abramov and co-workers [14].

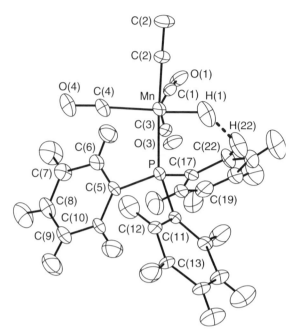

**Figure 4.7**   Molecular structure of monohydride HMn(CO)$_4$PPh$_3$, shown with 60% probability ellipsoids based on neutron refinement. (Reproduced with permission from ref. 14.)

Figure 4.7 depicts the structure obtained for this molecule by low-temperature neutron and high-resolution x-ray diffraction using a charge-coupled-device area detector. According to the structural data, the Mn monohydride does not show unusually short intermolecular interactions, whereas a short *intramolecular* contact is clearly observed between the ortho hydrogen atom in one of the phenyl rings and the hydride atom MnH. The Mn–H···H–C distance is measured as 2.101 Å, which is smaller than the sum of the van der Waals radii of H but remains quite long. The molecular geometry of the system shows similar C–H···H and H···H–Mn angles, determined as 129.0 and 126.5°, respectively, due to the intramolecular character of the interaction.

The experimental effective charges found in the crystal show that the hydride ligand, H(1), is negatively charged (−0.40 e) and the ortho phenyl hydrogen, H(22), exhibits a positive charge of +0.32 e. The electrostatic component of the M–H···H–C interaction has been obtained in terms of electrostatic multipole moments found from the refined atomic charge density. This electrostatic contribution to the hydrogen bond energy is calculated as 5.7 kcal/mol. Finally, the experimental topological analysis shows that the $\rho_C$ and $\nabla^2\rho_C$ parameters are determined in the H···H direction as 0.066 and 0.79 au, respectively. In full accord with AIM theory, the small electron density and the positive Laplacian strongly support the formulation of this interaction as dihydrogen bonding.

## 4.2. GAS-PHASE EXPERIMENTS WITH DIHYDROGEN-BONDED COMPLEXES

Whereas theoretical studies of dihydrogen bonding where dihydrogen-bonded complexes are isolated, approximating the gas phase, are very numerous, gas-phase experiments with these complexes are relatively rare. This circumstance is probably connected with the nontrivial experimental techniques used for gas-phase investigations.

The first data on gas-phase complexes formed between strong acids and the methane molecule were reported by Barnes in 1983 [15]. The author observed a $CH_4 \cdot HCl$ complex in a rare gas matrix that was assigned to a van der Waals system. Its $C_{3v}$ structure has also been confirmed by gas-phase microwave spectroscopy [16]. Other complexes, such as $CH_4 \cdot HCN$, $CH_4 \cdot HF$, and $CH_4 \cdot HBr$, were reported later [17]. Since these complexes represent $C-H \cdots H$ interactions, their nature is discussed in Chapter 6.

Since the $\nu(O-H)$ and $\nu(N-H)$ stretching vibrations in IR spectra of proton donors are very sensitive to the formation of classical hydrogen bonds, these vibrations have been probed by Patwari and co-workers [18] in gas mixtures containing phenol and borane dimethylamine (BDMA). The IR spectra have been recorded by the fluorescence-detected IR (FDIR) technique developed by Tanabe and co-workers [19]. In the framework of this approach, the population of a particular species in the ground state is monitored by laser-induced fluorescence (LIF) with UV laser excitation, and a change in its intensity is recorded as a function of the frequency of the IR laser, introduced 50 ns prior to the UV laser. When the IR frequency is resonant with the vibrational transition of the species, the ground-state population decreases and leads to a depletion in the LIF signal. Previously, the authors examined electronic transitions of the phenol–BDMA complex by the LIF excitation spectrum of phenol and its clusters with BDMA. The systems were prepared in a supersonic free jet by an expansion of the gaseous mixture of phenol and BDMA seeded in helium buffer gas at 3 atm. The sample was heated to 310 K to obtain sufficient vapor pressure of BDMA.

As shown in Figure 4.8(A), the FDIR spectrum of phenol exhibits O–H stretching vibration at 3657 $cm^{-1}$ corresponding to the monomer phenol molecules. The region 3200 to 3700 $cm^{-1}$ in the FDIR spectrum recorded for the phenol–BDMA complex is shown in Figure 4.8(B). Here the band detected at 3272 $cm^{-1}$ belongs to the N–H stretching vibration of the BDMA moiety. However, phenol O–H stretching vibration is now detected at 3483 $cm^{-1}$. Thus, the band undergoes a low-frequency shift by 174 $cm^{-1}$ typical of hydrogen bonding with phenol acting as a strong proton donor. Similar spectral data shown in Figure 4.9 have been collected for a complex formed by phenol and BTMA. However, in this case an analysis of B–H stretching vibrations in the region 2300 to 2460 $cm^{-1}$ (Figure 4.10) has led to determination of the proton-accepting site. The individual phenol molecules are not active in this region (A), whereas the phenol–BTMA cluster shows a sharp band at 2438 $cm^{-1}$ and two broadened bands at 2394 and 2405 $cm^{-1}$. The $BH_3$ group in free BTMA has a $C_3$ local

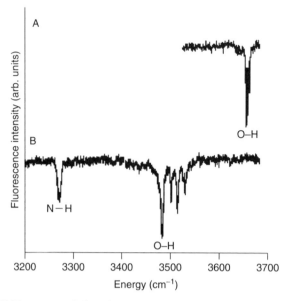

**Figure 4.8**    FDIR spectra of phenol (A) and the phenol–BDMA complex (B). (Reproduced with permission from ref. 18.)

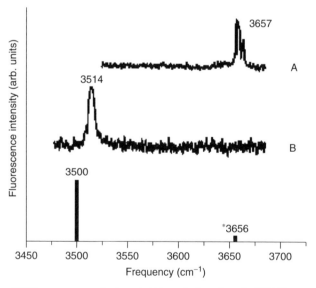

**Figure 4.9**    FDIR spectrum of phenol (A) and the phenol–BTMA complex (B) in the O–H stretching region. The bottom lines correspond to the theoretical calculations performed for the complex depicted in Figure 4.11 and the phenol monomer, marked with an asterisk. (Reproduced with permission from ref. 20.)

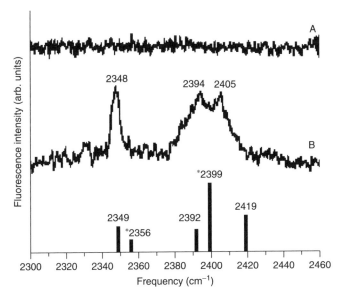

**Figure 4.10** FDIR spectrum of phenol (A) and the phenol–BTMA complex (B) in the B–H stretching region. The bottom lines correspond to the calculations carried out for the complex shown in Figure 4.11 and the BTMA monomer, marked with an asterisk. (Reproduced with permission from ref. 20.)

symmetry that decreases at complexation. Therefore, the appearance of three distinct peaks indicates clearly that the BH$_3$ group in BTMA acts as a proton acceptor to yield the dihydrogen O–H$\cdots$H–B cluster. It is important that the experimental spectroscopic data have been supported by calculations of the IR frequencies performed for the complex, whose geometry has been obtained at the B3LYP/6-311++G level (Figure 4.11).

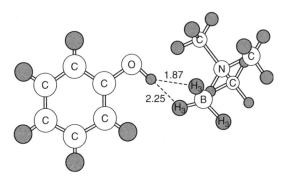

**Figure 4.11** Geometry of the phenol–BTMA complex optimized at the B3LYP/ 6-311++G level. The marked distances are measured in angstroms. (Reproduced with permission from ref. 20.)

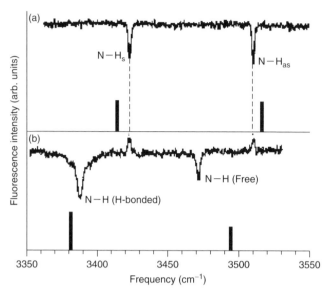

**Figure 4.12**    FDIR spectra of (a) aniline and (b) a dihydrogen-bonded aniline–BTMA complex. The bottom lines correspond to the spectra calculated. (Reproduced with permission from ref. 21.)

Similar spectroscopic data have been reported for a gaseous dihydrogen-bonded complex formed by aniline with BTMA [21]. The spectral features in the region of the N–H stretching vibrations are most interesting and therefore are shown in Figure 4.12(a). The FDIR spectrum of isolated aniline exhibits two bands of the NH$_2$ group, observed at 3422 and 3509 cm$^{-1}$. The bands can be attributed to symmetric and asymmetric N–H stretching vibrations. The N–H bands of the aniline–BTMA complex are shown in Figure 4.12(b), where the broad band at 3387 cm$^{-1}$ belongs to the hydrogen-bonded N–H group. At the same time, the higher-frequency band can be assigned to the free N–H bond. However, the N–H frequency shifts observed at complexation are not straightforward as in the case of the phenol O–H band. The NH$_2$ group of free aniline, having a $C_{2v}$ local symmetry, corresponds to two identical but *coupled* N–H bonds. As a result it gives lower-frequency *symmetric* and higher-frequency *asymmetric* stretching vibrations. When the complex is formed, the symmetry is lowering and both N–H bonds become decoupled. The free N–H stretching shows a value (3475 cm$^{-1}$, median of $\nu_S$ and $\nu_{AS}$) which is very close to that detected in the aniline–BTMA complex (3471 cm$^{-1}$). Therefore, the difference between the free and dihydrogen-bonded stretching N–H frequencies is calculated as 84 cm$^{-1}$.

Finally, independent evidence for the existence of dihydrogen bonds in the gas phase has been obtained by mass spectroscopy [22] applied to the dihydrogen-bonded phenol–BTMA complex. In addition, the study of fragmentation of ions in the phenol–BTMA complex has revealed molecular hydrogen elimination in the gas phase.

## 4.3. EXPERIMENTS WITH DIHYDROGEN-BONDED COMPLEXES IN SOLUTIONS

In contrast to solid or gas phases, the formation of dihydrogen complexes in solutions is always a reversible equilibrium:

$$n(\text{X--H})_a + m(\text{H--Y})_b \rightleftharpoons n(\text{X--H})_a \cdots m(\text{H--Y})_b \qquad (4.1)$$

where $n$ and $m$ symbolize the stoichiometry of complexation, and $a$ and $b$ show self-association. Rates of direct and opposite reactions, usually diffusion controlled, depend on the nature of proton donors and proton acceptors, their concentrations, and the temperature and properties of solvents. It is also obvious that the stoichiometry of the complexation can be dependent on concentrations and temperature.

Among various physicochemical methods, IR spectroscopy and $^1$H NMR are most appropriate tools for the study of dihydrogen bonds in solution. However, it is worth mentioning that these methods are basically different. First, they measure physical properties that change upon complexation: bond vibrations and magnetic behavior. Second, equilibrium (4.1) is usually slow on the IR spectroscopy time scale and very fast on the NMR time scale. In other words, proton donors, proton acceptor, and their complexes are detected separately in IR spectra, whereas the NMR parameters of these moieties are usually averaged.

### 4.3.1. IR Spectral Criteria for the Formation of Dihydrogen-Bonded Complexes in Solutions

Since the stretching frequency of a chemical bond can be approximated by Hooke's law,

$$v = \frac{1}{2\pi} \left( \frac{k}{m} \right)^{1/2} \qquad (4.2)$$

where $v$ is the frequency of the vibration, $k$ the force constant, and $m$ the mass, changes in bond lengths should be accompanied by corresponding frequency shifts in IR spectra. The appearance of these shifts does not depend on the aggregate state of compounds investigated. Thus, the $v(\text{XH})$ and $v(\text{HY})$ shifts observed for proton donors and proton acceptors can serve as the IR criteria for dihydrogen bond formation. These criteria, summarized by Shubina and co-workers [23], are based primarily on the IR behavior of proton-donor components HOR. It should be noted that they are very similar to those formulated for classical hydrogen bonds.

According to these criteria, a dihydrogen bond is formed when in the presence of a proton acceptor, the $v(\text{OH})$ band intensity of a free proton-donor component decreases and a new low-frequency broad $v(\text{OH})$ band (part of the dihydrogen-bond complex) appears. As in the case of classical hydrogen bonds,

**TABLE 4.1. IR Spectral Parameters [$v$ (OH) Regions] Obtained for Free and Dihydrogen-Bonded Proton Donors in $CH_2Cl_2$ Solutions of the Ions $[B_{10}H_{10}]^{2-}$ and $[B_{12}H_{12}]^{2-}$**

| Hydride | Proton Donor | $v$(OH) free $(cm^{-1})$ | $v$(OH) bound $(cm^{-1})$ | $\Delta v$(OH) $(cm^{-1})$ |
|---|---|---|---|---|
| $[B_{10}H_{10}]^{2-}$ | $Pr^iOH$ | 3607 | 3501 | 106 |
| | EtOH | 3611 | 3498 | 113 |
| | MeOH | 3624 | 3500 | 124 |
| | $CF_3CH_2OH$ | 3601 | 3440 | 161 |
| | $(CF_3)_2CHOH$ | 3580 | 3368 | 212 |
| | PhOH | 3586 | 3405 | 181 |
| | $4\text{-}FC_6H_4OH$ | 3586 | 3398 | 188 |
| | $4\text{-}NO_2C_6H_4OH$ | 3562 | 3327 | 235 |
| | $(CF_3)_3COH$ | 3520 | 3300 | 220 |
| $[B_{12}H_{12}]^{2-}$ | $Pr^iOH$ | 3607 | 3529 | 78 |
| | EtOH | 3611 | 3528 | 83 |
| | $CF_3CH_2OH$ | 3601 | 3490 | 111 |
| | $(CF_3)_2CHOH$ | 3580 | 3422 | 158 |
| | PhOH | 3586 | 3450 | 136 |
| | $4\text{-}FC_6H_4OH$ | 3586 | 3448 | 137 |
| | $4\text{-}NO_2C_6H_4OH$ | 3562 | 3394 | 168 |
| | $(CF_3)_3COH$ | 3520 | 3357 | 163 |

IR experiments should be performed in the absence of self-association and specific solvates. These conditions can be reached using low concentrations of proton donors and inert solvents such as hexane and $CH_2Cl_2$.

IR spectral changes depend on the nature of proton donors; they are more pronounced in the presence of stronger proton donors. Table 4.1 illustrates these observations in solutions of decahydro-*closo*-decaborate and dodecahydro-*closo*-dodecaborate anions $[B_{10}H_{10}]^{2-}$ and $[B_{12}H_{12}]^{2-}$ (**I** and **II** in Figure 4.13), hydrogen atoms of which act as proton-accepting sites [25]. It is clear that the $\Delta v$(OH)

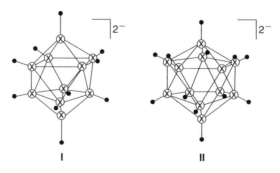

**Figure 4.13**

shifts are significant but minimal ($\sim$100 cm$^{-1}$) for relatively weak proton donors such as Pr$^i$OH, and maximal (>200 cm$^{-1}$) for stronger acids.

In contrast to the obvious spectral behavior of proton donors, determination of proton-accepting sites by IR spectra is not easy and is particularly difficult if a proton acceptor has a number of potential proton-accepting centers. Anion [B$_{10}$H$_{10}$]$^{2-}$ containing apical and equatorial B–H bonds illustrates this situation well. Since the B–H stretching for B–H bonds that participate in dihydrogen bonding could be low-frequency shifted and the free B–H bonds could undergo high-frequency shifts, these effects produce a complex combination. It follows from Figure 4.14 that an addition of (CF$_3$)$_2$CHOH to [B$_{10}$H$_{10}$]$^{2-}$ leads to a broad, slightly low-frequency-shifted asymmetric $\nu$(BH) band (spectrum 2). In such cases it is very useful to perform the experiments with selectively deuterated anions [25].

The IR spectra of anion [1,10-B$_{10}$H$_8$D$_2$]$^{2-}$, containing D in the equatorial positions, are shown in Figure 4.15. The spectrum of free [1,10-B$_{10}$H$_8$D$_2$]$^{2-}$ exhibits a $\nu$(BH) band at 2450 cm$^{-1}$, while a $\nu$(BD) band is observed at 1887 cm$^{-1}$ (curve 1) corresponding to a regular isotopic ratio of 1.32. Addition of (CF$_3$)$_2$CHOH leads to decreasing the $\nu$(BD) intensity and the appearance of a broad low-frequency $\nu$(BD) band which can be assigned to the dihydrogen-bonded adduct. Thus, only apical hydrogen atoms accept the protons and high-frequency bands $\nu$(BD) (1892 cm$^{-1}$), and $\nu$(BH) (2467 cm$^{-1}$) can be assigned to stretching vibrations of the terminal B–D and B–H bonds in BD(H)$\cdots$HOR complexes. The variable-temperature IR experiments support these assignments. As seen, the intensity of the $\nu$(BD) band, corresponding to the dihydrogen-bonded complexes, remarkably increases on cooling form 250 to 200 K.

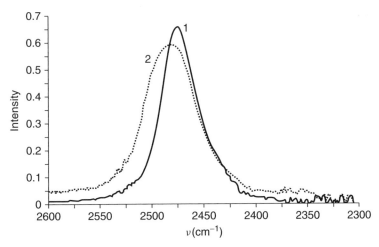

**Figure 4.14**    The 200 K IR spectra in the B–H stretching region recorded in a CH$_2$Cl$_2$ solution of [B$_{10}$H$_{10}$]$^{2-}$ (1) and in the presence of (CF$_3$)$_2$CHOH (2). (Reproduced with permission from ref. 25.)

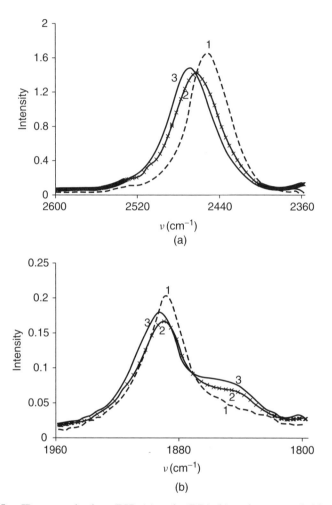

**Figure 4.15** IR spectra in the $\nu(BH)$ (a) and $\nu(BD)$ (b) regions recorded in a $CH_2Cl_2$ solution of anion $[1,10\text{-}B_{10}H_8D_2]^{2-}$ (1) and in the presence of $(CF_3)_2CHOH$ at 200 K (2) and 250 K (3). (Reproduced with permission from ref. 25.)

Similarly, in the case of transition metal hydrides acting as proton acceptors in dihydrogen-bonded complexes, spectral changes in the $\nu(MH)$ regions are not simple. For example, interaction of monohydride $WH(CO)_2(NO)(PEt_3)_2$ [26] with $(CF_3)_2CHOH$ results in the appearance of a new low-frequency $\nu(WH)$ *shoulder* ($\approx$ at 1640 cm$^{-1}$) which is shifted by only 30 cm$^{-1}$ with respect to the initial $\nu(WH)$ band observed at 1670 cm$^{-1}$. Dihydride $[MeC(CH_2PPh_2)_3RuH_2$ $(CO)]$ also shows insignificant changes: The $\nu(RuH)$ band undergoes a shift from 1871 cm$^{-1}$ to 1851 cm$^{-1}$ even in the presence of strong proton donors. It is obvious that such data should be supported by independent experiments: for example, by $^1H$ NMR spectra, which are more sensitive to dihydrogen bonding.

**TABLE 4.2. Equilibrium Constants Characterizing Intermolecular Associations Between a Proton Donor and a Proton Acceptor**[a]

| [Me$_3$NBH$_3$] (mol/L) | $K$ (L/mol) PhOH/Me$_3$ NBH$_3$ = 1 : 1 | $K$ (L$^2$/mol$^2$) PhOH/Me$_3$ NBH$_3$ = 1 : 2 | $K$ (L$^2$/mol$^2$) PhOH/Me$_3$ NBH$_3$ = 2 : 1 |
|---|---|---|---|
| 0 | — | — | — |
| 0.1 | 5.1 | 69 | 117 |
| 0.2 | 4.8 | 31 | 148 |
| 0.3 | 5.1 | 21 | 200 |
| 0.4 | 5.1 | 15 | 245 |
| 0.5 | 5.0 | 12 | 281 |
| 0.6 | 5.0 | 9 | 317 |

[a]$K$ for the proton donor, PhOH, and the proton acceptore, Me$_3$NBH$_3$, were obtained at 25°C in CCl$_4$ solutions at different concentrations of Me$_3$NBH$_3$ by assuming different stoichiometry.

### 4.3.2. How to Determine the Stoichiometry of Dihydrogen-Bonded Complexes in Solution by IR Spectroscopy

The stoichiometry of dihydrogen-bonded complexes in solution is generally unknown. Since the IR stretching vibrations of proton-donor components are very sensitive to dihydrogen bonding, they can be probed at different concentrations of proton acceptors to determine the composition of complexes. Brown and co-workers [27] give a good example of such a determination performed for PhOH interacting with Me$_3$NBH$_3$ in inert CCl$_4$. The free-OH absorbance can be taken as a measure of the concentration for the isolated phenol molecule. Knowledge of this parameter together with the known initial concentrations of PhOH and Me$_3$NBH$_3$ allow us to calculate equilibrium constants, $K$, assuming various compositions of the complexes (Table 4.2). As shown, the $K$ magnitude remains constant only at 1 : 1 stoichiometry but not for 1 : 2 or 2 : 1 complexes.

### 4.3.3. Energy Parameters of Dihydrogen-Bonded Complexes from IR Spectra in Solution

As in the case of any equilibrium, the enthalpy ($\Delta H^\circ$) and entropy ($\Delta S^\circ$) parameters can be obtained for dihydrogen bonding on the basis of variable-temperature IR spectra recorded in the stretching–vibration regions of proton donors or proton acceptors. Since the IR spectroscopy is "a fast" physical method, free and bound moieties are observed separately. Thus, variable-temperature IR spectra can be used for determination of the corresponding equilibrium constants at each temperature. Then the $\Delta H^\circ$ and $\Delta S^\circ$ values can be obtained easily via the van't Hoff equation. It is obvious that if the indicator bands are not overlapping, IR experiments lead to the most accurate data. This situation takes place, for example, in the case of the CO group in the Ru dihydride interacting with (CF$_3$)$_2$CHOH:

$$(\text{triphos})\text{Ru(CO)H}_2 + \text{HOCH(CF}_3)_2 \rightleftharpoons (\text{triphos})\text{Ru(CO)(H)H} \cdots \text{HOCH(CF}_3)_2$$

$$(4.3)$$

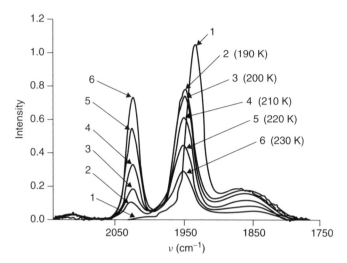

**Figure 4.16**    Variable-temperature IR spectra in the $\nu(CO)$ region of dihydride [(triphos)Ru(CO)H$_2$] in CH$_2$Cl$_2$ at 190 K (1) and in the presence of a 10-fold excess of (CF$_3$)$_2$CHOH recorded between 190 and 230 K (curves 2 to 6). (Reproduced with permission from ref. 28.)

Figure 4.16 shows variable-temperature IR spectra recorded in the $\nu$ (CO) region for a CH$_2$Cl$_2$ solution of this dihydride upon addition of a 10-fold excess of (CF$_3$)$_2$CHOH. The temperature evolution of the $\nu$ (CO) intensities in such spectra is analyzed easily to give the equilibrium constants. The pattern in Figure 4.17 shows the temperature dependence of these constants leading to $-\Delta H^\circ$ and $-\Delta S^\circ$ values of 6.7 kcal/mol and 19.6 eu, respectively.

Except for the direct van't Hoff method, the enthalpies of dihydrogen bonds can be estimated from the frequency shifts, $\Delta\nu$ (HX), or the integral intensity, $A$ (XH), determined in the IR spectra for proton-donor components, HX:

$$-\Delta H^\circ = \begin{cases} \dfrac{18\Delta\nu(HX)}{720 + \Delta\nu(HX)} & (4.4) \\[2mm] 0.30[\Delta\nu(HX)]^{1/2} & (4.5) \\[2mm] 2.9(\Delta A)^{1/2} & (4.6) \end{cases}$$

It is worth mentioning that initially, these correlations were found for classical hydrogen-bonded complexes by Iogansen [24] on the basis of a large mass of IR data treated statistically. Relationship (4.4) is more common and applicable in the very large frequency range between 0 and 2000 cm$^{-1}$. Correlation (4.5) is rather limited by the $\Delta\nu$ values, which are greater than 200 cm$^{-1}$. Finally, the relationship (4.6) is the best. In fact, according to experience, this relationship works for different hydrogen-bonded systems forming in solution, the solid state, or the gas phase, enthalpies of which are between 0.1 and 15 kcal/mol. However,

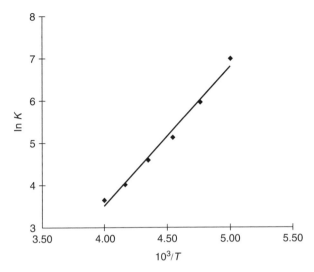

**Figure 4.17**   Temperature dependence of the equilibrium constant (4.3) determined from IR spectra. (Reproduced with permission from ref. 28.)

since measurements of the $\Delta v$(HX) parameters are methodically simpler, the relationship (4.4) is used more often.

As has been shown experimentally, the relationships 4.4–4.6 can be used successfully for dihydrogen-bonded complexes [24]. In addition, despite their approximate character, they lead to enthalpy values that are usually close to those determined by van't Hoff's method. For example, applying eq. (4.4) to equilibrium equation (4.3) gives a $-\Delta H^\circ$ value of 6.6 kcal/mol versus the 6.7 kcal/mol found using van't Hoff's method.

### 4.3.4. $^1$H Nuclear Magnetic Resonance Evidence for Dihydrogen Bonding in Solution

In the framework of $^1$H NMR spectroscopy, the formation of dihydrogen bonds is connected with changes in electron environments of $^1$H nuclei that participate in dihydrogen bonding. In other words, dihydrogen bonding can be seen from the $^1$H chemical shifts of both proton donors and proton acceptors.

Theoretically, the NMR chemical shift ($\delta$) is a sum of diamagnetic and paramagnetic terms and the terms caused by the magnetic anisotropy of neighboring groups and polarity of solvents [29]. The diamagnetic term connects with the electron environment, whereas the paramagnetic contribution is affected by the energy of electron excitation (for details, readers are referred to any NMR text). A combination of these terms is not trivial and therefore the chemical shifts observed do not reflect directly the electronic density around the target nuclei. Despite this circumstance, $\delta$ values and their changes upon formation of dihydrogen bonds, particularly, can be used phenomenologically to formulate $^1$H NMR criteria.

$^1$H NMR spectroscopy is a slow method on the time scale of dihydrogen bond formation. Even at very low temperatures, resonances in the $^1$H NMR spectra are averaged between dihydrogen-bonded complexes and isolated proton-donor and proton-acceptor molecules. In some cases, use of Freon solvents and the low-temperature technique still allows us to stop dihydrogen bond formation on the $^1$H NMR time scale. For example, the ReH resonance of hydride Cp*ReH(CO) (NO) decoalesces in the presence of acidic alcohols at 96 K to give two well-resolved lines at −7.54 and −8.87 ppm attributed to the free and dihydrogen-bonded ReH resonances, respectively [30]. The difference between the free and coordinated states is significant (≈1.3 ppm) and provides reliable detection of dihydrogen bonding in solutions.

Similar to classical hydrogen bonds formed by acids and regular organic bases, the $^1$H resonances of proton-donor components undergo significant down-field displacements (>2 ppm) in the presence of proton acceptors. For example, the OH resonance of $(CF_3)_2CHOH$ shifts from 2.726 ppm to 4.690 ppm in a $C_6D_{12}$ solution even in the presence of the relatively weak proton acceptor $BH_3NEt_3$ [31]. Thus, despite the average character of the chemical shifts, they can be used as criteria for dihydrogen bond formation.

As we have shown above, determination of proton-acceptor centers by IR spectra is not simple, and a similar situation can take place in $^1$H NMR experiments. For example, the chemical shift difference observed for the B–$^1$H resonance of $BH_3NEt_3$ in the absence and presence of $(CF_3)_2CHOH$ is very small, even with an excess of the alcohol. In fact, the $^1H\{^{11}B\}$ resonance of $BH_3NEt_3$ at 1.532 ppm has been found to appear as a narrow line (due to $^{11}B$ decoupling) in a $C_6D_{12}$ solution, transforming to a strongly broadened resonance centered at 1.490 ppm in the presence of $(CF_3)_2CHOH$. Thus, the B–H resonance undergoes a small *high-field displacement* of 0.042 ppm upon dihydrogen bonding. At the same time, high-field shifts are remarkably stronger for transition metal hydrides, acting as proton acceptors bonding to acids HX. Generally, these high-field displacements comprise −0.2 to −2.0 ppm and are direct evidence that hydridic hydrogens accept protons.

Let us demonstrate $^1$H NMR applications in practice. As mentioned above, MeH resonances are usually averaged between free and dihydrogen-bonded states. Therefore, their chemical shifts depend on both HX concentration and temperature, as is clear from the variable-temperature $^1$H NMR spectra of the dihydrogen-bonded system in eq. (4.3). A low-temperature addition of a twofold excess of $(CF_3)_2CHOH$ to a $CD_2Cl_2$ solution of the Ru dihydride leads to a *temperature-dependent* high-field shift of the hydride resonance from −7.32 ppm to −7.86 ppm (see curve 2 in Figure 4.18). In contrast, the chemical shift of individual Ru dihydride is practically independent of the temperature (curve 1). At 200 K, curve 2 reaches a plateau (−7.86 ppm), showing clearly that the equilibrium (4.3) is shifted completely to the dihydrogen-bonded adduct. Since the hydride resonance in the presence of $(CF_3)_2CHOH$ even at 200 K is still averaged between two magnetically nonequivalent but equally populated states (i.e.,

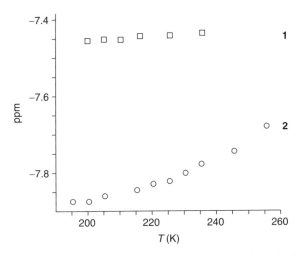

**Figure 4.18** Temperature dependence of the chemical shift of the hydride resonance of (triphos)Ru(CO)H$_2$ in CD$_2$Cl$_2$ (squares) and in the presence of a twofold excess of (CF$_3$)$_2$CHOH (circles). (Reproduced with permission from ref. 28.)

the free and bonded hydride ligands), the chemical shift of the hydride ligand involved in dihydrogen bonding can be calculated as $-8.29$ ppm via the equation

$$\delta(\text{RuH})^{\text{obs}} = 0.5\delta(\text{RuH})^{\text{free}} + 0.5\delta(\text{RuH}\cdots\text{HOR}) \tag{4.7}$$

This example demonstrates practically how to determine the chemical shift of dihydrogen-bonded complex from averaged NMR parameters.

One of the important advantages of $^1$H NMR spectroscopy is that it makes possible calculation of the H$\cdots$H distances in dihydrogen-bonded complexes formed in solution. As we have shown above, the formation of dihydrogen bonds is accompanied by the appearance of H$\cdots$H contacts that are shorter than a sum of the van der Waals radii of H ($< 2.4$ Å). Such short H$\cdots$H contacts are connected directly with the spin–lattice ($T_1$) $^1$H NMR relaxation of target nuclei [32]. The contacts between two $^1$H nuclei cause strong homonuclear dipolar coupling, DC$_{\text{H–H}}$, which is written as

$$\text{DC}_{\text{H–H}} = 0.3 \left(\frac{\mu_0}{4\pi}\right)^2 \gamma_{\text{H}}^4 \hbar^2 r(\text{H–H})^{-6} \tag{4.8}$$

where $\mu_0$, $\gamma_{\text{H}}$, and $\hbar$ are well-known constants and $r(\text{H–H})$ is the distance between the $^1$H nuclei. In turn, strong dipolar coupling results in an additional contribution to nuclear dipole–dipole relaxation:

$$\frac{1}{T_1(\text{H}\cdots\text{H})} = \text{DC}_{\text{H–H}} \left(\frac{\tau_c}{1 + \omega_{\text{H}}^2 \tau_c^2} + \frac{4\tau_c}{1 + 4\omega_{\text{H}}^2 \tau_c^2}\right) \tag{4.9}$$

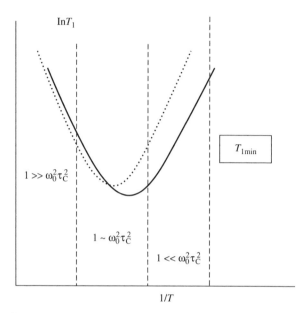

**Figure 4.19** Temperature dependence of the dipolar $T_1$ relaxation time in semi logarithmic coordinates at the Larmor frequency $\omega_0$. The dashed $T_1$ curve corresponds to a higher Larmor frequency. The regions with $1 >> \omega_0^2\tau_C^2$, $1 \sim \omega_0^2\tau_C^2$, and $1 << \omega_0^2\tau_C^2$ correspond to fast, intermediate, and slow molecular motions on the frequency scale of NMR.

Since the molecular motion correlation time, $\tau_c$, depends on the temperature, the $T_1$ curves, plotted in semilogarithmic coordinates, go through a minimum, $T_{1,min}$ (Figure 4.19) which can be expressed as

$$T_{1,min}(\text{H–H}) = \left[\frac{r(\text{H–H})}{5.815}\right]^6 \nu \qquad (4.10)$$

where $T_{1,min}$, $r(\text{H–H})$, and the working frequency of a NMR spectrometer, $\nu$, are measured in seconds, Å, and MHz, respectively. Thus, the shortened $^1\text{H}$ $T_1$ time measured for the $^1\text{H}$ resonance of a proton acceptor site will itself be good evidence of the formation of dihydrogen bonds.

Since determinations of the H$\cdots$H distance are not trivial and should be based on the $^1\text{H}$ $T_1$ relaxation contribution caused by a single H$\cdots$H contact, corresponding to a dihydrogen bond, let us illustrate this approach by an example of the Ru dihydride in eq. (4.3) [28]. The variable-temperature $^1\text{H}$ $T_1$ data collected for the hydride resonance in the presence of $(\text{CF}_3)_2\text{CHOH}$ have shown a $^1\text{H}$ $T_1$ minimum ($^1\text{H}$ $T_{1,min}^{\,obs}$) of 0.119 s at 200 K, characterizing the dihydrogen-bonded adduct, $(\text{triphos})\text{Ru(CO)(H)H}\cdots\text{HOCH(CF}_3)_2$ (see the plateau in Figure 4.18). Since the hydride observed resonance still undergoes a fast exchange between the free and bounded RuH states, the $T_{1,\text{min}}$ magnitude observed can be expressed as

$$\frac{1}{T_{1,\text{min}}^{\text{obs}}} = \frac{0.5}{0.178} + \frac{0.5}{T_{1,\text{min}}^{\text{obs}}(\text{RuH}\cdots\text{H})} \tag{4.11}$$

where 0.178 s has been measured as the minimal $^1$H $T_1$ time of the individual Ru dihydride [in the absence of $(\text{CF}_3)_2\text{CHOH}$], and the $T_{1,\text{min}}^{\text{obs}}(\text{RuH}\cdots\text{H})$ is a minimal relaxation time of the hydride ligand that is involved in dihydrogen bonding. The $T_{1,\text{min}}^{\text{obs}}(\text{RuH}\cdots\text{H})$ value is calculated from eq. (4.11) as 0.0894 s. This time is remarkably shorter than that in the individual dihydride (0.178 s), due to an additional hydride–proton dipolar coupling. Then this additional relaxation rate, $1/T_{1,\text{min}}(\text{RuH}\cdots\text{H})$, governed by the single hydride–proton dipolar contact, is expressed as

$$\frac{1}{T_{1,\text{min}}(\text{RuH}\cdots\text{H})} = \frac{1}{T_{1,\text{min}}^{\text{obs}}(\text{RuH}\cdots\text{H})} - \frac{1}{0.178} \tag{4.12}$$

and calculated as 0.183 s. This value can be used in eq. (4.10) to determine the $r(\text{H}\cdots\text{H})$ distance as 1.81 Å. Despite the fact that the $^1$H $T_1$ times of the proton-acceptor sites depend on many factors, dihydrogen bonding generally leads to reducing $T_1$ by a factor 1.5 to 2. The latter can be accepted as the most important NMR criterion. It should be added that the temperature-dependent effects [nuclear overhauser effects (NOEs) relative to relaxation phenomena and observed for the resonance of proton donors (OH) and proton acceptors (metal-H), can also be used in the diagnosis of dihydrogen-bonded complexes [26].

Finally, a very sensitive tool for the diagnosis of dihydrogen bonds in solution has been suggested by Alyllon and co-workers [33]. However this tool is not universally applicable because it is based on the exchange H–H coupling observed in some transition metal hydrides, such as in Structure 4.2. The $J(\text{H}_a-\text{H}_b)$ splitting in these systems has a quantum mechanical origin and can reach abnormally large values. As has been demonstrated, the formation of dihydrogen-bonded complexes causes a dramatic increase in exchange coupling values: for example, to 249 Hz in the complex shown in Structure 4.2 in the presence of $(\text{CF}_3)_2\text{CHOH}$ versus 74 Hz observed in the free hydride molecule.

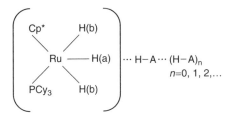

**Structure 4.2**

### 4.3.5. Energy Parameters of Dihydrogen Bonds in Solution from $^1$H NMR

Since the nature of the hydride chemical shifts, particularly in transition metal hydride complexes, is not simple [32], there is no reliable correlation between $\delta_H$ and the enthalpy of dihydrogen bonding. Nevertheless, the chemical shifts of hydride resonances and their changes with temperature and the concentration of proton-donor components, for example, can be used to obtain the energy parameters for dihydrogen bonding in solution. As earlier, the enthalpy ($\Delta H^\circ$) and entropy ($\Delta S^\circ$) values can be obtained on the basis of equilibrium constants determined at different temperatures. Let us demonstrate some examples of such determinations.

The hydride chemical shifts for free and bounded RuH ligands in dihydride [(triphos)Ru(CO)H$_2$] in eq. (4.3) were determined in Section 4.3.4. Curve 2 in Figure 4.18 is a direct way to calculate the molar fractions of free and bound components and thus to determine the equilibrium constants at different temperatures. These constants yield $-\Delta H^\circ$ and $-\Delta S^\circ$ values of 7.1 kcal/mol and 19.0 eu, respectively, in good agreement with IR data.

A more complicated situation, worthy of separate discussion, has been observed by Shubina and co-workers [26] in the $^1$H NMR spectra of a toluene-d$_8$ solution of WH(CO)$_2$(NO)(PMe$_3$)$_2$ in the presence of (CF$_3$)$_2$CHOH. Figure 4.20 illustrates the temperature effects observed. The dependencies are clearly pronounced,

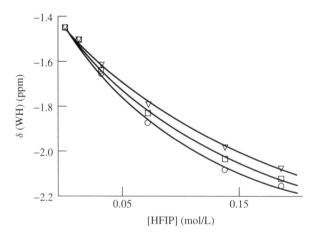

**Figure 4.20**   Chemical shift of the hydride resonance in a toluene-d$_8$ solution of hydride WH(CO)$_2$(NO)(PMe$_3$)$_2$ in the presence of (CF$_3$)$_2$CHOH as a function of temperature and proton-donor concentration: $-70^\circ$C (triangles), $-80^\circ$C (squares), and $-90^\circ$C (circles). (Reproduced with permission from ref. 26.)

**Structure 4.3**

but they do not show a plateau. In other words, under these conditions, the equilibrium

$$WH(CO)_2(NO)(PMe_3)_2 + HOCH(CF_3)_2 \rightleftharpoons W(CO)_2(NO)(PMe_3)_2$$

$$H \cdots HOCH(CF_3)_2 \tag{4.13}$$

is not completely shifted toward the dihydrogen-bonded adduct, and thus the chemical shift of its hydride resonance remains unknown. Assuming the formation of a $1:1$ dihydrogen-bonded adduct in equilibrium 4.13, the chemical shift observed, $\delta_{WH}$, can be expressed as

$$\delta_{WH} = (\delta_F + 0.5\delta_B) \left[ \frac{(a^2 + 4KW)^{1/2} - a}{1 + 0.5(a^2 + 4KX)^{1/2} - a} \right] \tag{4.14}$$

where $a = KW - KX + 1$, $\delta_F$ and $\delta_B$ are chemical shifts of the free and bounded transition metal complex, $K$ is the equilibrium constant, and W and X are initial concentrations of the W hydride and the proton donor, respectively. The curves in Figure 4.20 can be treated by computer fittings to determine $K$. The equilibrium constants have been determined from 6.3 L/mol at $-70°$ C to 10.8 L/mol at $-90°$ C) and the $\delta_B$ value has been obtained as $-2.67$ ppm. Then calculations of energy parameters of dihydrogen bonding are not problematic.

An original NMR method for the determination of dihydrogen bond strengths has been used by Peris and co-workers [34] for the intramolecular dihydrogen-bonded iridium complexes in Structure 4.3. The authors have suggested the barrier of the C–N rotation, occurring on the NMR time scale, as a measure of bonding strength in the intramolecular dihydrogen-bonded iridium complexes shown. In turn, this barrier can be determined from a line-shape analysis of the magnetically nonequivalent NH$_2$ resonances observed in temperature-dependent $^1$H NMR spectra. In fact, the rotational barrier represents a difference between energies of the ground state, where the Ir complexes are stabilized by intramolecular dihydrogen bonds, and the transition state, where these bonds are absent. Therefore, the energy values obtained by the NMR spectra associate with relative magnitudes

**TABLE 4.3. Strength of Intramolecular Dihydrogen Bonding in Iridium Hydride Complexes (Structure 4.3) as a Function of the Nature of the *trans*-Ligand Y**

| Trans-Ligand Y | Energy (kcal/mol) |
| --- | --- |
| H | 5.0 |
| CO | 3.7 |
| CN | 3.4 |
| I | 3.3 |
| MeCN | 3.1 |
| Br | 3.0 |
| Cl | 2.9 |
| F | < 2.9 |

that characterize $H \cdots H$ bonding at variation in the nature of the *trans*-ligand Y. The absolute energy values have been determined by independent studies of C–N rotation in systems without dihydrogen bonds. As follows from Table 4.3, intramolecular dihydrogen bonds change from weak ($< 2.9$ kcal/mol) to medium (5.0 kcal/mol) as a function of Y.

## 4.4. CONCLUDING REMARKS

1. High-quality x-ray or neutron diffraction is the best method for visualization of dihydrogen bonds in the solid state and determination of $H \cdots H$ distances $< 2.4$ Å as a criterion of dihydrogen bonding. X-ray diffraction experiments with charge density lead to direct characterization of bonding between two hydrogens in the framework of topological analysis of the electronic density.

2. The IR spectra recorded by the fluorescence detection technique, where the population of a species in the ground state is monitored by laser-induced fluorescence with UV laser excitation, can be used successfully to detect dihydrogen bonds in the gas phase. It has been found, for example, that the $\nu(O–H)$ and $\nu(N–H)$ stretching vibrations in such IR experiments are very sensitive to dihydrogen bonding.

3. IR spectroscopy is the best method for detection of dihydrogen bonds in solution. The low-frequency shifts of stretching-vibration bands observed for proton donors [e.g., $\nu$ (OH)], are a very good criterion for the formation of dihydrogen bonds. It has been found that these shifts are very similar for dihydrogen and hydrogen bonding.

4. Dihydrogen bonding energies can be determined in solution by variable-temperature IR spectra in the framework of van't Hoff's method or from the frequency shifts, $\Delta\nu$ (HX) [or integral intensity, $\Delta A$(XH)], determined in the IR spectra of proton-donor components, which correlate with the bonding energy.

5. The resonances of proton-donor and proton-acceptor components in $^1H$ NMR spectra are also useful for the detection of dihydrogen bonds. Low- and high-field shifts observed for proton donors and proton acceptors, respectively, clearly show dihydrogen bonding in solution. The most important advantage of $^1H$ NMR spectroscopy is the possibility of calculating the $H \cdots H$ distances in solutions of dihydrogen-bonded complexes. The calculations are based on $^1H$ $T_1$ relaxation experiments. The enthalpy and entropy of dihydrogen bonding can be obtained on the basis of equilibrium constants determined by NMR spectra at various temperatures.

## REFERENCES

1. G. R. Desiraju, *J. Chem. Soc. Dalton Trans.* (2000), 3745.
2. T. Steiner, G. R. Desiraju, *Chem. Commun.* (1998), 891.
3. W. H. Zachariasen, R. C. L. Mooney, *J. Chem. Phys.* (1934), **2**, 34.
4. S. Marincean, R. Custelcean, R. S. Stein, J. E. Jackson, *Inorg. Chem.* (2005), **44**, 45.
5. L. Brammer, W. T. Klooster, F. R. Lempke, *Organometallics* (1996), **15**, 1721.
6. V. I. Bakhmutov, E. V. Vorontsov, *Rev. Inorg. Chem.* (1998), **18**, 183.
7. W. T. Klooster, T. F. Koetzle, P. E. M. Siebahn, T. B. Richardson, R. H. Crabtree, *J. Am. Chem. Soc.* (1999), **121**, 6337.
8. T. B. Richardson, S. De Gala, R. H. Crabtree, P. E. M. Siegbahn, *J. Am. Chem. Soc.* (1995), **117**, 12875.
9. R. Custelcean, J. E. Jackson, *Chem. Rev.* (2001), **101**, 1963.
10. E. S. Shubina, I. A. Tikhonova, E. V. Bakhmutova, F. M. Dolgushin, M. Y. Antipin, V. I. Bakhmutov, I. B. Sivaev, L. N. Teplitskaya, I. T. Chizhevsky, I. V. Pisareva, V. I. Bregadze, L. M. Epstein, V. B. Shur, *Chem. Eur. J.* (2001), **7**, 3783.
11. R. H. Crabtree, *Acc. Chem. Res.* (1996), **29**, 348.
12. T. Raminal, H. Jong, I. D. McKenzie, M. Jenning, J. A. C. Clyburne, *Chem. Commun.* (2003), 1722.
13. T. S. Koritsanszki, P. Coppens, *Chem. Rev.* (2001), **101**, 1583.
14. Y. A. Abramov, L. Brammer, W. T. Klooster, R. M. Bullock, *Inorg. Chem.* (1998), **37**, 6317.
15. A. J. Barnes, *J. Mol. Struct.* (1983), **100**, 259.
16. A. C. Legon, R. P Roberts, A. I. Wallrock, *Chem. Phys. Lett.* (1990), **173**, 107.
17. M. J. Atkins, A. C. Legon, A. I. Wallrock, *Chem. Phys. Lett.* (1992), **192**, 368.
18. G. N. Patwari, A. Fujii, N. Mikami, *J. Chem. Phys.* (2000), **113**, 9885.
19. S. Tanabe, T. Ebata, M. Fujii, N. Mikami, *Chem. Phys. Lett.* (1993), **215**, 347.
20. G. N. Patwari, A. Fujii, N. Mikami, *J. Chem. Phys.* (2006), **124**, 241103.
21. G. N. Patwari, T. Ebata, N. Mikami, *Chem. Phys.* (2002), **283**, 193.
22. G. N. Patwari, T. Ebata, N. Mikami, *J. Phys. Chem. A* (2001), **105**, 10753.
23. N. V. Belkova, E. S. Shubina, L. M. Epstein, *Acc. Chem. Res.* (2005), **38**, 624.
24. L. M. Epstein, E. S. Shubina, *Coord. Chem. Rev.* (2002), **231**, 165.

25. E. S. Shubina, E. V. Bakhmutova, A. M. Filin, I. B. Sivaev, L. N. Teplitskaya, A. L. Chistaykov, I. V. Stankevich, V. I. Bakhmutov, V. I. Bregadze, L. M. Epstein, *J. Organomet. Chem.* (2002), **657**, 155.

26. E. S. Shubina, N. V. Belkova, A. N. Krylov, E. V. Vorontsov, L. M. Epstein, D. G. Gusev, M. Niedermann, H. Berke, *J. Am. Chem. Soc.* (1996), **118**, 1105.

27. M. P. Brown, R. W. Heseltine, P. A. Smith, P. J. Walker, *J. Chem. Soc. A* (1970), 410.

28. V. I. Bakhmutov, E. V. Bakhmutova, N. V. Belkova, C. Bianchini, L. M. Epstein, D. Masi, M. Peruzzini, E. S. Shubina, E. V. Vorontsov, F. Zanobini, *Can. J. Chem.* (2001), **79**, 478.

29. R. K. Harris, *Nuclear Magnetic Resonance Spectroscopy: A Physicochemical View*, Longman Scientific & Technical, Harlow, England (1986).

30. E. S. Shubina, N. V. Belkova, A. V. Ionidis, N. S. Golubev, L. M. Epstein, *Russ. Chem. Bull.* (1997), **44**, 1349.

31. L. M. Epstein, E. S. Shubina, E. V. Bakhmutova, L. N. Saitkulova, V. I. Bakhmutov, A. I. Chistyakov, I. V. Stankevich, *Inorg. Chem.* (1998), **37**, 3015.

32. V. I. Bakhmutov, *Practical Nuclear Magnetic Resonance Relaxation for Chemists*, Wiley, Hoboken, NJ (2005).

33. J. A. Alyllon, S. Sabo-Etienne, B. Chaudret, S. Ulrich, H. H. Limbach, *Inorg. Chim. Acta* (1997), **259**, 1.

34. E. Peris, J. C. Lee, J. R. Rambo, O. Eisenstein, R. H. Crabtree, *J. Am. Chem. Soc.* (1995), **117**, 3485.

# 5

# INTRAMOLECULAR DIHYDROGEN BONDS: THEORY AND EXPERIMENT

Intramolecular interaction is a powerful factor that controls molecular architecture, particularly in the case of geometrically flexible molecular systems. The existence and energies of intramolecular classical hydrogen bonds and their role in chemistry and biochemistry are well known. They stabilize molecular conformations, promote short- and long-range proton transfers, participate in the creation of three-dimensional structures of large molecules and play a fundamental role in the phenomenon of molecular recognition.

If a molecule contains groups with hydrogen atoms, one of which is acidic and the second atom is able to accept the proton, this molecule is potentially capable of intramolecular dihydrogen bonding. Such a geometrical situation takes place, for example, in the molecule of the alcohol in Structure 5.1 containing the relatively strong proton-donor site O–H and the weak proton-acceptor site C–H. It is obvious that an energetically simple rotation around a single C–O bond creates a geometrical ground for the formation of a dihydrogen bond that could stabilize the closed conformation of this molecule. Intuitively, it is clear that such bonds will be quite weak. Nevertheless, they exist and contribute to structures of alcohols, amino acids, and even molecules with weakly polarized C–H bonds. It should be added that the stronger intramolecular proton–hydride interactions in transition metal hydride systems are important intermediates in the heterolytic splitting of dihydrogen.

One important subject considered in this chapter is establishment of criteria that allow us to express molecular systems containing short intramolecular $H\cdots H$ contacts as systems with intramolecular dihydrogen bonding. In this context, theoretical approaches based on AIM theory are particularly accurate. For example, geometrical features of molecules that have weak or strong proton-donor and proton-acceptor sites can only provoke close $H^{+\delta}\cdots^{-\delta}H$ contacts, due to simple electrostatic interactions. However, interaction energies can be negligibly small. In this connection it is important to distinguish weak electrostatic attraction from dihydrogen bonding.

*Dihydrogen Bonds: Principles, Experiments, and Applications*, By Vladimir I. Bakhmutov
Copyright © 2008 John Wiley & Sons, Inc.

**Structure 5.1**

In addition, we discuss various physicochemical methods, such as x-ray or neutron diffraction, [1]H NMR, [1]H NMR relaxation, and IR spectroscopy, used to detect intramolecular dihydrogen bonds and their characterizations. We begin with very weak interactions and progress to systems containing stronger H···H bonds, with bonding energies reaching 5 kcal/mol and more.

## 5.1. WEAK INTRAMOLECULAR BONDING: C–H···H–C IN SYSTEMS WITH SLIGHTLY POLARIZED BONDS CH

On the basis of numerous theoretical and experimental data, it is well know that transition metal polyhydride systems form a structural continuum in which extreme structures are terminal hydrides and $^2\eta-(H_2)$ complexes (Structure 5.2) [1]. If this idea is applied for carbon (in reality such an analogy is not valid chemically), H···H contacts between hydrogen atoms attached to the same carbon atom (geminal hydrogens) could be a simple example of intramolecular dihydrogen interaction C–H···H–C (but not dihydrogen bonds). These hypothetical interactions, being very weak, could contribute to total enthalpies of formation, for example, of hydrocarbons. In reality there are no experimental or theoretical evidence for the existence of this geminal interaction. Nevertheless, it is surprising, but additions of geminal H–H terms to empirical additive bond energy schemes lead to better predictions of the formation energies for compounds of this class [2].

Principally, short nongeminal C–H···H–C contacts (intramolecular and intermolecular as well) can be found in many organic molecules, particularly when C–H bonds are slightly polarized. As we show below, the strength of H···H interaction decreases strongly with increasing H···H separation. On the basis of this correlation, a small polarization of C–H bonds does not provide H···H contacts smaller than 2 Å [3]. Then, from a conservative point of view, the H···H interactions will be a continuous mix of coulombic and dipolar interactions.

**Structure 5.2**

$R_2C$—H

Ph P

Ph

H

**Structure 5.3**

As to the background of these weak interactions, steric compression by bulky groups (e.g., in the compound shown in Structure 5.3) or packing forces may contribute significantly to the close H···H contacts. In fact, a CSD search performed for organic compounds similar to Structure 5.3, has revealed the presence of Ph–H···H–C contacts smaller than 2.4 Å in 61 molecules, and 17 contacts were even smaller than 2.2 Å [4]. It is obvious that in the absence of additional data (specifically, the electron density topology in the H···H directions), treatment of the C–H···H–C contacts in terms of dihydrogen bonding will be doubtful despite the shortened distances determined from x-ray structures.

Strong experimental and theoretical evidence for the existence of intramolecular dihydrogen bonding C–H···H–C has been obtained by Grabowski and co-workers [5], who reported the x-ray single-crystal structures of [4-((E)-but-1-enyl)-2,6-dimthoxyphenyl]pyridine-3-carboxylate (BDMP) and [4-((E)-pent-1-enyl)-2,6-dimthoxyphenyl]pyridine-3-carboxylate (PDMP), depicted in Figure 5.1, where the intramolecular C–H···H–C contacts, shown as dashed lines, have been determined as 2.20 and 2.17 Å for BDMP and PDMP, respectively. Both of these distances are even smaller than a sum of the van der Waals radii of H, 2.4 Å, and thus the close contacts can be formulated formally as dihydrogen bonds. These surprising cases have been tested in the framework of AIM theory, where molecular geometries have been optimized at the B3LYP/6-311++G$^{**}$ level. The calculations led to molecular graphs, one of which is shown in Figure 5.2. This graph demonstrates clearly the presence of a bond critical point on the C–H···H–C direction, and thus the H···H interactions found in such x-ray structures can actually be assigned to dihydrogen bonds. According to the calculations, the energies of dihydrogen bonds were in the range 2 to 3 kcal/mol, classifying these interactions as weak.

To extend the investigations and the AIM tests, the authors have carried out high-level ab initio calculations of simpler models such as styrene and its fluoro derivatives. The treatments have shown that intramolecular dihydrogen bonds such as those described above are not unique, and for example, the contour density map in Figure 5.3 obtained for the styrene molecule again exhibits a bond critical point in the H···H direction with a distance of 2.208 Å. In addition, the topological analysis of the electron density has shown a small $\rho_C$ value (0.0089 au) and a positive Laplacian, $\nabla^2\rho_C$, of 0.0419 au, both typical of hydrogen bonding. In addition, planar conformations of these molecules had the lowest energies, revealing the stabilizing character of dihydrogen bonding. However, let

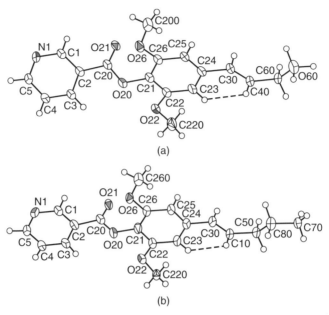

(a)

(b)

**Figure 5.1** Crystal structure of (a) [4-((E)-but-1-enyl)-2.6-dimthoxyphenyl]pyridine-3-carboxylate (BDMP) and (b) [4-((E)-pent-1-enyl)-2.6-dimthoxyphenyl]pyridine-3-carboxylate (PDMP) with the intramolecular C–H···H–C contacts 2.20 and 2.17 Å, respectively. (Reproduced with permission from ref. 5.)

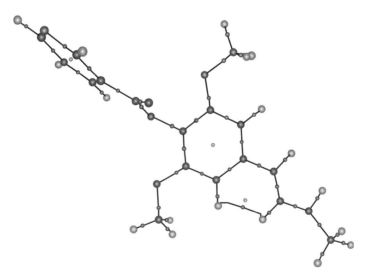

**Figure 5.2** Molecular graph of [4-((E)-but-1-enyl)-2.6-dimthoxyphenyl]pyridine-3-carboxylate with bond critical and ring critical points. (Reproduced with permission from ref. 5.)

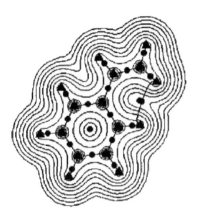

**Figure 5.3** Contour density map obtained for the styrene molecule at a B3LYP/ 6-311++G** level. (Reproduced with permission from ref. 5.)

us state a reservation here: Because of the absence of C–H polarizations, Bader termed such interaction H–H bonding or H–H interaction (see Section 2.1) but not dihydrogen bonding (see below).

## 5.2. INTRAMOLECULAR DIHYDROGEN BONDS IN SOLID AMINO ACIDS: C–H BONDS AS WEAK PROTON ACCEPTORS

Since classical hydrogen bonds perform a fundamental function in life science, Pakiari and Jamshidi [6] have focused on amino acids as molecules potentially capable of intramolecular dihydrogen bonding. In fact, they contain the groups OH and NH as relatively strong proton donors and C–H bonds that could play the role of weak proton acceptors. In this connection, three amino acids—glycine, alanine, and proline—have been probed theoretically at the B3LYP/6-311G** level following the requirements suggested for intramolecular dihydrogen bonds:

1. The formation of an intramolecular dihydrogen bond should be accompanied geometrically by the creation of a ring of size four to six, with the five-membered cycle preferred.
2. The dihydrogen bond formation should lead to the stabilization energy $E_{db}$, obtained as the difference between energies of molecules with and without dihydrogen bonds.
3. Dihydrogen bonding should lead to the same molecular conformation as in the absence of H···H interactions.

It is worth mentioning that point 3 is particularly important for correct estimations of bonding energies.

Figure 5.4 depicts dihydrogen bonds C–H···HO and C–H···H–N that were found in structurally optimized amino acids, and in Table 5.1 we list their

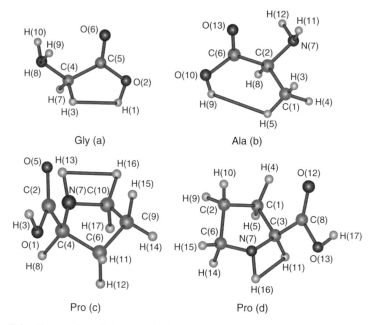

**Figure 5.4** Geometries of glycine, alanine, and proline optimized at a B3LYP/6-311G** level. The solid lines show intramolecular dihydrogen bonds. (Reproduced with permission from ref. 6.)

**TABLE 5.1. Geometric Parameters of the Amino Acids Shown in Figure 5.4, Optimized Without and with Dihydrogen Bonding and Corresponding Stabilization Energies of Dihydrogen-Bonded Systems**

| Amino Acid | $H \cdots H$ (Å) | $H_{db} \cdots H_{db}$ (Å) | Dihydral Angle $X-H \cdots H-Y$ (deg) | Dihydral Angle $X-H_{db} \cdots H_{db}-Y$ (deg) | $E_{db}$ (kcal/mol) |
|---|---|---|---|---|---|
| Gly (a) | 2.565 | 2.365 | 103.2 | 121.8 | −2.1 |
| Ala (b) | 2.588 | 2.354 | 115.7 | 126.2 | −1.4 |
| Pro (c) | 2.533 | 2.252 | 118.1 | 153.6 | −2.5 |
| Pro (d) | 2.455 | 2.333 | 131.1 | 154.8 | −0.7 |

geometrical and energy parameters. In accordance with point 3, the conformations of molecules investigated do not actually change upon dihydrogen bonding beyond dihedral angles and $H \cdots H$ distances, which increase and decrease, respectively. It follows from Table 5.1 that stabilization energy $E_{db}$ depends remarkably on the nature of acids but takes *nonnegligible* values in all cases. Independently, these data have been confirmed by replacement of carbon atoms by silicon atoms, which have less electronegativity.

**Structure 5.4**

Similar to amino acids, O–H···H–C contacts in closed conformation to 2-cyclopropyl ethenol (Structure 5.4), and its F-derivatives, have been analyzed by Palusiak and Grabowski [7]. It has been found that these molecules exhibit bifurcated dihydrogen bonds formed by C–H proton acceptors. Calculations performed at the MP2/6-311++G(d,p) and B3LYP/6-311++G(d,p) levels resulted in H···H contacts of 2.0 to 2.2 Å, with the bifurcated H···H distances in each molecule being practically identical. Thus, these distances are again shorter than the sum of the van der Waals radii of H. However, it is very important that the open conformations obtained by rotation around C–O bonds were energetically *more stable*. In addition, all the net charges on hydrogen atoms were positive. These data classify the H···H contacts rather as van der Waals interactions. At the same time, as has been noted by Palusiak and Grabowski, the closed conformations show slightly elongated O–H and C=C bonds. This fact corresponds rather closely to H···H bonding, but the bonding is very weak. The latter is supported by detailed AIM analysis of the systems, including the topology of the electronic density. It follows from Table 5.2 that the bond critical points found in the H···H directions show $\rho_C$ values within the range 0.0067 to 0.0112 au, with positive Laplacian values of 0.0250 to 0.0416 au, typical of weak bonding.

**TABLE 5.2. H···H Distances and Topological Electron Density Parameters for Closed Conformations of the Compounds in Structure 5.4[a]**

| $R_1$ | $R_2$ | H···H (Å) | $\rho_C$ (au) | $\nabla^2\rho_C$ (au) |
|-------|-------|-----------|---------------|------------------------|
| H | H | 2.017 | 0.0111 | 0.0413 |
|   |   | 2.100 | 0.0092 | 0.0336 |
| F | H | 2.137 | 0.0086 | 0.0329 |
|   |   | 2.234 | 0.0070 | 0.0261 |
| H | F | 2.013 | 0.0112 | 0.0416 |
|   |   | 2.100 | 0.0092 | 0.0334 |
| F | F | 2.151 | 0.0083 | 0.0317 |
|   |   | 2.254 | 0.0067 | 0.0250 |

[a]Calculated at the MP2/6-311++G(d,p) (upper values) and B3LYP/6-311++G(d,p) (lower values) levels.

This is a very good illustration, supporting the idea that there is no clear border between van der Waals interactions and weak dihydrogen bonds.

## 5.3. INTRAMOLECULAR DIHYDROGEN BONDS: C–H···H–B

In Section 5.2 C–H bonds played the role of weak proton-accepting sites. In contrast, the $8,9'$-[*closo*-[3-Co($\eta^5$-C$_5$H$_5$)]-1,2-C$_2$B$_9$H$_{10}$]$_2$ dimer shows an x-ray structure with two close C–H···H–B contacts, where C–H bonds evidently act as proton donors (see Figure 5.5). On the basis of the x-ray data it has been suggested that the solid-state conformation in Figure 5.5 is stabilized by a pair of dihydrogen bonds with distances C(6)H···H–B(8′) and C(7)–H···H–B(12′), determined as 2.097 and 1.996 Å, respectively. Both distances are actually significantly smaller than 2.4 Å. The C–H···H (114.48 and 151.12°) and H···H–B (138.89 and 110.39°) angles are also in a good agreement with this suggestion.

One of the important criteria for dihydrogen bonding is the loss of charge of hydrogen atoms participating in the interaction. This tendency is observed clearly in calculations of atomic charges. The charges on atoms H(6), H(7), H(8′), and H(12′) in the dimer structure have been calculated as 0.254, 0.258, −0.022, and 0.01 e, respectively, versus 0.260, 0.260, 0.003, and 0.033 for the same atoms in a monomer structure.

For geometrical reasons, multiple intramolecular C–H···B–H contacts can be expected in the aminoboron hydride shown in Structure 5.5 and also in relative compounds. In fact, these compounds are very stable. It is probable that the intramolecular dihydrogen bonds could act against their own disproportion [9].

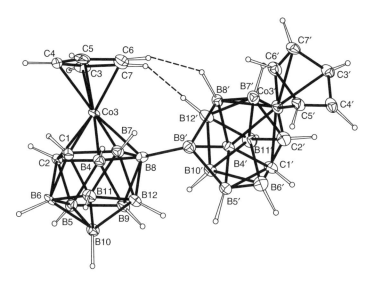

**Figure 5.5** X-ray single-crystal structure of $8,9'$-[*closo*-[3-Co($\eta^5$-C$_5$H$_5$)]-1,2-C$_2$B$_9$H$_{10}$]$_2$. The dashed lines show dihydrogen bonds. (Reproduced with permission from ref. 8.)

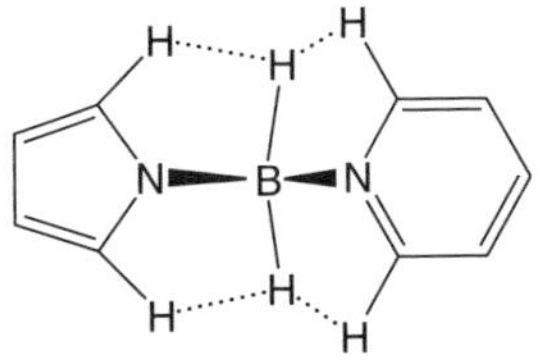

**Structure 5.5**

On the one hand, x-ray crystallography data could support this idea because the heterocycles are coplanar relative to the B–H bonds. Moreover, in the case of the compound in Structure 5.5, this conformation remains in solution, according to $^1$H NOE NMR measurements. On the other hand, the solid-state CH···HB distances are too long ($\leq$ 2.65 Å), and the HF/6-31G* calculations carried out for the compound have shown, that where the heterocyclic rings are oriented orthogonally, the conformation is more stable. Thus, this question is still open. It is quite probable that the interaction between protonic hydrogens on the $\alpha$-carbons and hydridic BH hydrogens is electrostatically attractive.

Figure 5.6 shows azacyclopentane- and azacyclohexane-borane adducts where groups $BH_3$ or $BF_3$ constantly occupy the equatorial positions. These conformations have been found in the solid state by x-ray diffraction [10] as well as by $^1$H NMR and $^1$H nuclear Overhauser enhancement spectroscopy (NOESY) NMR

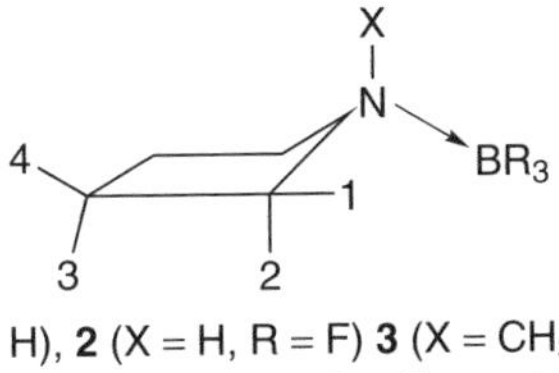

**1** (X = R = H), **2** (X = H, R = F) **3** (X = CH₃, R = H), **4** (X = CH₃, R = F)

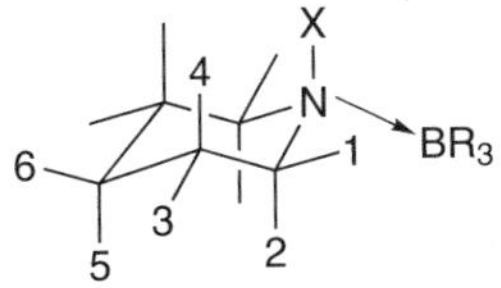

**5** (X = R = H), **6** (X = H, R = F) **7** (X = CH₃, R = F),

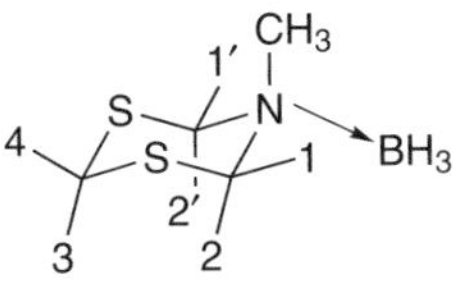

**8**

**Figure 5.6**

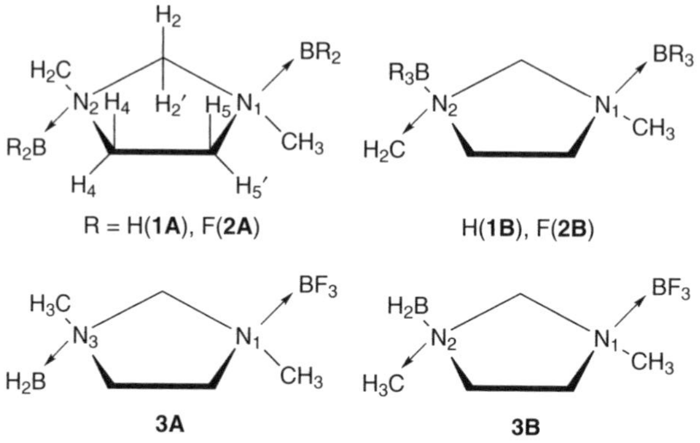

Figure 5.7

experiments in solution [11]. Again, intramolecular interaction C–H$\cdots$H–B or C–H$\cdots$F–B, involving the protons of $\alpha$-CH$_2$ groups and hydrogen or fluorine atoms in the BH$_3$ or BF$_3$ groups could stabilize these structures. The $^1$H $T_1$ NMR relaxation measurements performed in solutions of molecules **1** and **8** led to CH$\cdots$H–B contacts of 2.30 and 2.12 Å, smaller or close to the sum of the van der Waals radii of H. Similar distances are observed in the solid state.

Since intramolecular contacts C–H$\cdots$H–B in cyclic aminoboranes are still elongated, they could represent a lowest limit of weak interactions. In addition, the space feature of the BH$_3$ groups itself could provoke the appearance of such close contacts. Therefore, the nature of these interactions is still unclear. The answer came from DFT calculations at the B3LYP/6-31 + G(d,p) level and topological analysis of the electron density performed for molecules **1A, 1B, 2A, 2B, 3A, and 3B** in Figure 5.7 [12]. It has been found that twist molecular conformations in the gas phase are actually stabilized, due to C–H$\cdots$H–B or C–H$\cdots$F–B bonding. One molecular graph is shown in Figure 5.8, where the bond critical points are observed clearly in the C–H$\cdots$F–B directions. Since the $\rho_C$ values are small and the $\nabla^2\rho_C$ values are positive, the interactions are characterized as closed-shell. For example, the C–H$\cdots$H–B contacts in compound **1A** show $\rho_C$ and $\nabla^2\rho_C$ values of 0.0079 and 0.0220, respectively. Finally, these intramolecular interactions are weak because the energy of one proton–hydride/proton–fluoride contact was estimated as 1.9 and 0.7 kcal/mol, respectively. At the same time this energy seems to be sufficient to stabilize the conformational states.

## 5.4. INTRAMOLECULAR BONDS: N–H$\cdots$H–B AND O–H$\cdots$H–B

Since N–H and O–H bonds are obviously more polarized than C–H bonds, the corresponding dihydrogen bonds can be expected to be stronger. Nevertheless, examples of intramolecular bonding N–H$\cdots$H–B and O–H$\cdots$H–B are still not numerous. One can be found in recent computational modeling of oligomers

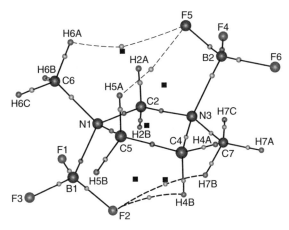

**Figure 5.8** Molecular graph of compound **2A** obtained at the B3LYP/6-31G+(d,p) level with the bond critical points noted as the smallest spheres and ring critical points noted as squares. (Reproduced with permission from ref. 12.)

$H(NH_2BH_2)_nH$, with $n$ from 1 to 6, interesting in the context of their potential applications for hydrogen storage [13]. Among different optimized chain structures, the closed conformations were most stable due to dihydrogen bonding, shown in Figure 5.9.

A second candidate for dihydrogen bonding of B–H···H–N is 2′-deoxycytidine-N(3)-cyanoborane, the molecular structure of which is depicted in Figure 5.10. The BH···HN distance in this molecule is 2.05 Å, smaller than the van der Waals sum, and the Mulliken charges at the hydrogen atoms in the BH···HN pair are opposite.

Excellent theoretical evidence for the existence of intramolecular dihydrogen bonds O–H···H–B has been reported by Grabowski [15]. The author carried out MP2 and HF calculations for the unsaturated alcohols in Structure 5.6 at different basis levels, where closed and opened conformation, geometries obtained by simple rotation around the C–O bond have been optimized and energy differences between these conformations have been taken as an energy measure of proton–hydride interactions. For example, MP2/6-311++G(3d,3p) calculations of compounds **1** and **2** gave strongly different but short H···H distances of 1.933 and 1.674 Å, which corresponded to dihydrogen bonding energy of −4.70 and −6.53 kcal/mol, respectively. The nature of these energetically medium dihydrogen bonds has been confirmed by topological analysis of the electronic densities. The bond critical points localized in the H···H directions have shown typical $\rho_C/\nabla^2\rho_C$ values of 0.016/0.048 and 0.027/0.063 au for compounds **1** and **2**, respectively. The results obtained from different basis sets were very similar.

Fundamentally important results have been obtained by calculations of the systems in Structure 5.6 at a large variation in the nature of R, strongly affecting bond lengths $d_1$, $d_2$, and $d_3$ and hence H···H distances [15,16]. These data

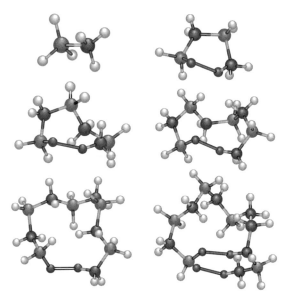

**Figure 5.9** Optimized structures of oligomers $H(BH_2NH_2)_nH$ with $n = 1$ to 6. The darker large spheres correspond to nitrogen atoms, and the dihydrogen bonds are shown as dark connections. (Reproduced with permission from ref. 13.)

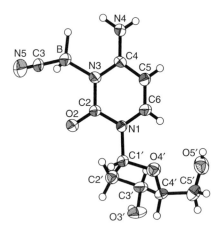

**Figure 5.10** Single-crystal x-ray structure of 2′-deoxycytidine-N(3)-cyanoborane with a BH···HN(4) distance of 2.05 Å. (Reproduced with permission from ref. 14.)

revealed a linear relationship between the bonding energy and the H···H bond length, shown in Figure 5.11. The dependence of the bonding energy on the electron density at bond critical points in the H···H directions has also been found to be linear (Figure 5.12). It is obvious that these relationships could be used to estimate electron density parameters when bonding energies are known.

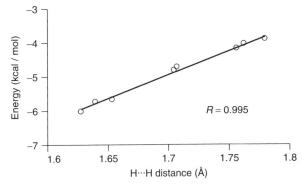

**Structure 5.6**   $R_1 = R_2 = R_3 = H$ **(1)**, $R_1 = Cl$, $R_2 = H$, $R_3 = Na$ **(2)**

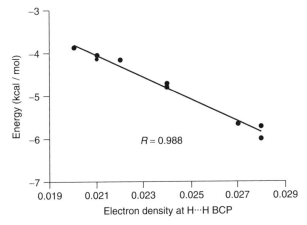

**Figure 5.11**   Linear relationship between the interaction energy and the H···H distances found for the intramolecular dihydrogen bonds O–H···H–B shown in Structure 5.6. (Reproduced with permission from ref. 16.)

**Figure 5.12**   Linear relationship between the interaction energy and the electron density at the H···H bond critical point found for the dihydrogen bonds shown in Structure 5.6. (Reproduced with permission from ref. 16.)

## 5.5. INTRAMOLECULAR DIHYDROGEN BONDS IN METAL HYDRIDE COMPLEXES

In 1994, Atwood and co-workers [17] reported on the x-ray structure of the organometallic molecule in Structure 5.7, where the strongly polarized bonds N–H and Al–H are potentially capable of dihydrogen bonding. In fact, the authors found a contact N–H$^{+\delta}\cdots^{-\delta}$H–Al of 2.31 Å, which is slightly smaller than 2.4 Å. However, for the high hydridic character of the hydrogen atoms bonded to Al, this distance seems to be too long and reflects, rather, the electrostatic character of the interaction, nevertheless stabilizing the eclipsed conformation. It is also probable that the electrostatic nature of the Al–H$\cdots$H–N interactions is responsible for $a \approx 3.0$ kcal/mol stabilization of a twist conformation in the trimer $(NH_2AlH_2)_3$ versus a chair conformation, shown in Figure 5.13. This result was shown by MP2 calculations.

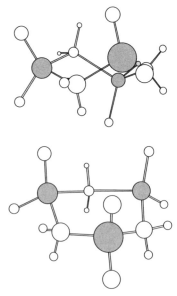

**Structure 5.7**

**Figure 5.13**  Twist and chair conformations of $(NH_2AlH_2)_3$. The darker spheres correspond to the Al atoms. (From ref. 18.)

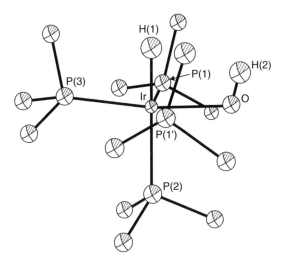

**Figure 5.14** Neutron diffraction structure of the transition metal hydride complex $(Me_3P)_4Ir(OH)H$. (Reproduced with permission from ref. 19.)

The first candidate for intramolecular interaction between proton-donor bond O–H and proton-acceptor bond M–H, where M is a transition metal, was reported in 1990 [19]. The neutron diffraction structure of iridium hydride $[(Me_3P)_4Ir(OH)H]$ clearly revealed the *syn*-oriented Ir–H(1) and O–H(2) bonds shown in Figure 5.14. All of the small atoms have been well localized to show a Ir–H···H–O distance of 2.4 Å. Again, the distance is too long for dihydrogen bonding and the interaction has a rather electrostatic character. Later, however, dihydrogen bonding Ir–H···H–N was established for the complexes shown in Structure 5.8. Crabtree and co-workers observed in $^1$H NMR spectra $^1J(H–H)$ *spin–spin coupling through a dihydrogen bond*: constants $^1J(H_{Ir}–H_a)$ were between 2 and 5 Hz, whereas no visible coupling was observed for $NH_b$ resonance [20].

The x-ray structure obtained for one of two possible isomers of the cationic iridium hydride complex [Structure 5.9(a)] exhibited two short N–H···H–Ir contacts of $1.75 \pm 0.05$ Å. These distances are significantly smaller than the sum

**Structure 5.8**

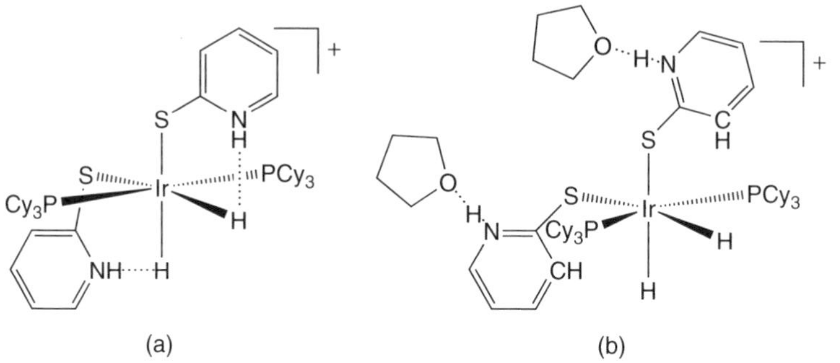

(a)                                      (b)

**Structure 5.9**

of the van der Waals radii of H. The same H···H separations have been measured in a $CD_2Cl_2$ solution by $^1H$ $T_{1,min}$ relaxation experiments. Independently, the short N–H···H–Ir contacts have been confirmed by $^1H$ NOE NMR spectra: Irradiation of the hydride resonance causes an 11% enhancement in the intensity of the NH line [21]. At such extremely short H···H distances, formulation of the N–H···H–Ir interactions as dihydrogen bonds seems to be obvious. On the other hand, the spin–spin coupling constant $^1J(^1H_{Ir}-^1HN)$ is not resolved in $^1H$ NMR spectra and is estimated as $\leq 2$ Hz only. It is probable that these H···H distances are underestimated and the bonding is not as strong. In fact, the dihydrogen bonds are easily destroyed when the iridium complex is placed into THF capable of intermolecular interaction with N–H protons [Structure 5.9(b)].

Short H···H distances have been measured by $^1H$ $T_{1,min}$ NMR relaxation in $CD_2Cl_2$ solutions of complex **9** ($1.82 \pm 0.05$ Å) in Structure 5.10 [22] and also for other closely related iridium complexes [23], where the distances were between $1.9 \pm 0.1$ and $2.1 \pm 0.1$ Å.

It is worth mentioning that the dihydrogen bond in solid complex **9** is elongated to 2.0 Å versus 1.82 Å in solution. According to the x-ray molecular structure, this effect correlates with the appearance of an additional contact between the NH proton and the $[BF_4]^-$ anion through the F atom due to a bifurcated bonding Ir–H···H(N)···F–B. The related iridium complexes have shown the same effects.

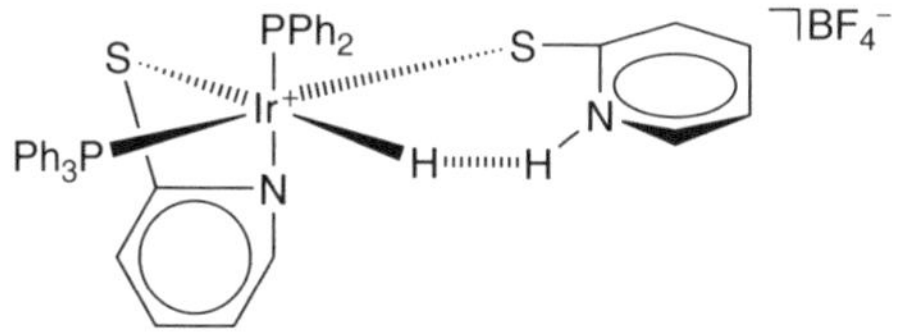

**Structure 5.10**

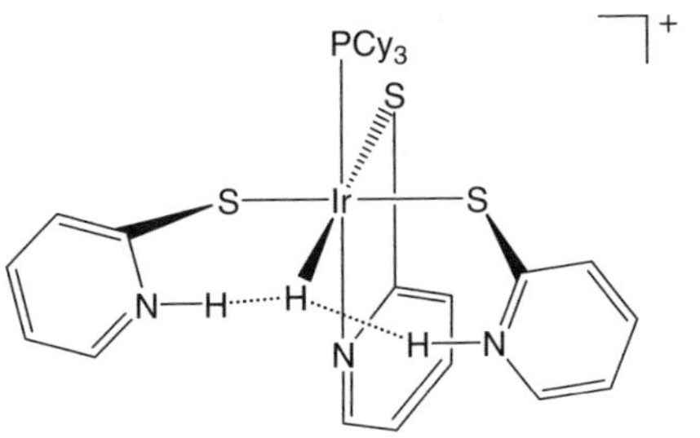

**Structure 5.11**

A unique double intramolecular Ir–H$\cdots$H–N contact has been found in the cationic iridium complex [IrH($\eta^1$-SC$_5$H$_4$NH)$_2$($\eta^2$-SC$_5$H$_4$NH)(PCy$_3$)][BF$_4$] shown in Structure 5.11, where an average H$\cdots$H distance was measured as 1.9 $\pm$ 0.2 Å in the solid state (the x-ray molecular structure) and 1.80 $\pm$ 0.03 Å in a CD$_2$Cl$_2$ solution ($^1$H $T_{1,min}$ measurements) [24]. Again, the H$\cdots$H contacts are shorter in solution.

Theoretical evidence for intramolecular dihydrogen bonds Ir–H$\cdots$H–X (X = O, N) has been obtained by investigations [25] of the neutral complex [Ir(H$_3$)(PH$_3$)(NHCH$_2$NH$_2$)] (compound **1** in Figure 5.15) and two posi-tively charged complexes, cis-[IrH(OH)(PH$_4$)][PF$_6$] and [IrH$_2$(CO)(PH$_3$)$_2$ (pzH–N)][BF$_4$]. Only the neutral complex has shown dihydrogen bonding, with H$\cdots$H bond lengths of 1.929 and 1.818 Å on the HF and MP2 levels, respectively. In accordance with the AIM criteria, the bond critical point is clearly observed in the H(9)–H(12) direction in Figure 5.16, where the $\rho_C$ and $\Delta^2\rho_C$ values are determined as 0.016 and 0.041 au, respectively. In contrast, the H$\cdots$H distance of $\sim$2 Å, formally smaller than 2.4 Å, obtained for the positively charged complex [IrH$_2$(CO)(PH$_3$)$_2$(pzH–N)][BF$_4$] by MP2 or HF calculations did not show the bond critical point in this H$\cdots$H direction. Thus, a short distance, alone, cannot be a criterion for the formation of dihydrogen bonds.

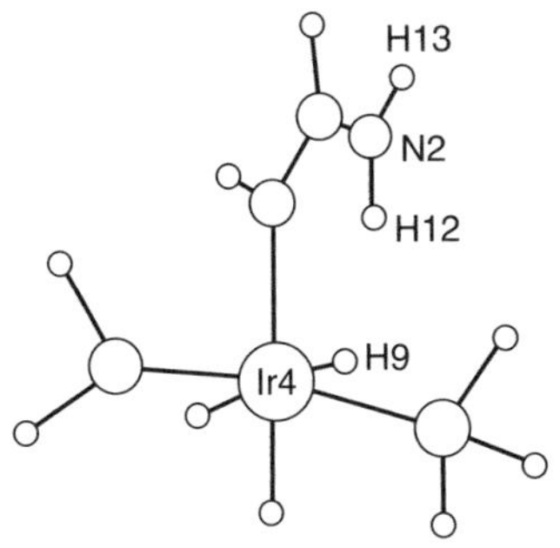

**1**

**Figure 5.15**

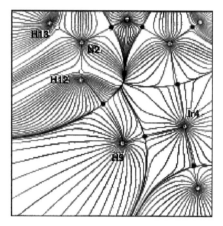

**Figure 5.16**   Gradients of the charge density, $\Delta\rho$, calculated for complex **1** in Figure 5.15 in the Ir–H···H–N plane. (Reproduced with permission from ref. 25.)

An intramolecular dihydrogen bond, Ru–H···H–N$^+$, in the charged Ru complex **1** in Scheme 5.1 could provoke the well-known proton transfer with the formation of Ru(H$_2$) complex **2** [26]. This bonding has been studied theoretically [27] to reveal the effect of the methylene bridge on the stability of model systems shown in Structure 5.12 to Structure 5.14, where the distances and the angles are given in angstroms and degrees, respectively. The geometries have

**Scheme 5.1**

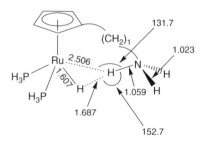

**Structure 5.12**

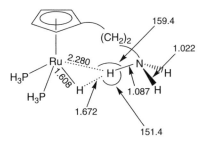

**Structure 5.13**

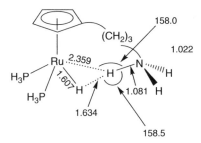

**Structure 5.14**

been optimized at the MP2/6-311++G(d,p) level, and the single-point energies have been calculated using the MP2 and B3LYP approaches at various basis sets. As shown, all of the systems exhibit H···H distances of 1.634 to 1.687 Å, significantly shorter than the sum of the van der Waals radii of H. A topological analysis has resulted in bond critical points in the H···H directions with the $\rho_C$ and $\Delta^2\rho_C$ values typical of dihydrogen bonding. Nevertheless, the geometry parameters look unusual. In fact, the shortest H···H distance in Structure 5.14 does not correspond to the longest N–H bond. In addition, the Ru···H–N angle in Structure 5.12 is closer to 180° than the H···H–N angle. Thus, the data show that the intramolecular Ru···H–N interactions cannot be ignored in such systems. In accordance with this statement, an energy analysis performed for these systems has demonstrated that the dihydrogen bonds contribute only 20% to the total stabilizing energy.

### 5.5.1. Intramolecular Dihydrogen Bonds in Metal Hydride Clusters

Few examples are known for this type of intramolecular dihydrogen bond. One of them is the trimetallic osmium cluster shown in Structure 5.15 [28]. This compound, described well by various methods, has revealed a number of problems connected with characterizations of dihydrogen-bonded complexes that deserve separate discussion. The hydrogen atoms localized in the x-ray molecular structure of complex **1** of Figure 5.17 provide a formulation of interaction N–H···

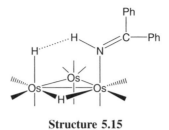

**Structure 5.15**

**Figure 5.17**  X-ray molecular structure of trimetallic osmium cluster **1**. (Reproduced with permission from ref. 28.)

H–Os as a dihydrogen bond because the H$\cdots$H distance is short [1.79(6) Å]. According to the linear relationship shown in Figure 5.11, this short distance corresponds to a quite strong bonding which could remain in inert solvents. However, the $^1$H $T_{1,\min}$ data, collected in a $CD_2Cl_2$ solution of **1**, resulted in a strongly elongated H$\cdots$H distance of $2.00 \pm 0.05$ Å. Since the difference is significant, it requires rationalization. As we have already noted, the x-ray diffraction method usually underestimates the H$\cdots$H distance. It is probably longer than 1.79(6) Å. In fact, when the positions for only atoms H(1), H(2), and H(3) were optimized by HF calculations at the fixed x-ray geometry of **1**, the H$\cdots$H distance became longer (1.89 Å). One can believe that this distance is the best estimation for the solid complex, but it is still short relative to the NMR data.

When the theoretical gas-phase model with complete geometry optimization was used for **1**, an H$\cdots$H distance of 1.93 Å resulted. This is probably the best approximation to a solution in $CD_2Cl_2$. All the data show a tendency in an *effective elongation* of the H$\cdots$H bond in going from the solid state to solution. This effect, which is probably common, can be explained by a large-amplitude oscillatory motion of the imine ligand along the N–Os coordination axis, which is effective in solutions.

The H$\cdots$H distance of the dihydrogen bond in **1**, corrected to 1.89 to 1.93 Å, could correspond to an interaction energy of $-3$ kcal/mol or less, again according to the correlation in Figure 5.11. It is interesting that similar bonding energy has been obtained experimentally via the IR spectra of complex **1** and the model osmium system, which is not capable of dihydrogen bonding. As shown in Figure 5.18, formation of the intramolecular dihydrogen bond is accompanied by the *red shift* typical of classical hydrogen bonding. Then, via relationship (4.4), this shift gives a $-\Delta H$ value of 2.25 kcal/mol, corresponding to rather weak dihydrogen bonds. In good agreement with this value, the B3LYP calculations have resulted in an interaction energy of $-2.6$ kcal/mol. In addition, the calculations have shown the opposite Mulliken charges at the hydrogen atoms, independently supporting the nature of the H$\cdots$H interaction.

In the cluster system above, the terminal hydridic hydrogen atom acts as a proton-accepting site. A triangular rhenium cluster (Figure 5.19) is a rare example of a *bridging* hydride ligand attracting a proton N–H [29]. It follows from the x-ray data that a solid cluster exists as a rotational conformer showing *syn*-orientation of H(b) and H–N with an H$\cdots$H distance of 2.24 Å. Formally, this distance is smaller than the van der Waals sum of 2.4 Å. However, in account for an inaccurate x-ray localization of hydrogen atoms in transition metal hydride

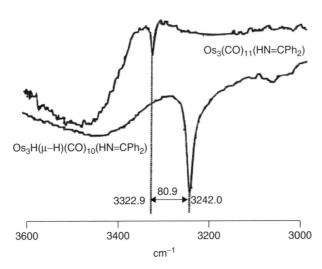

**Figure 5.18** IR spectra in the NH stretching region of cluster **1** with the dihydrogen bond and a model system $(HN{=}CPh_2)Os_3(CO)_{11}$. (Reproduced with permission from ref. 28.)

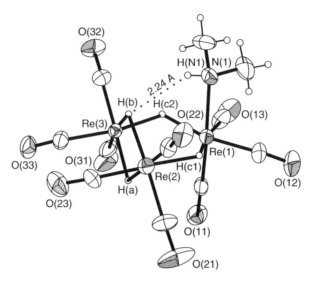

**Figure 5.19**   X-ray molecular structure of the $[Re_3(\mu\text{-}H)_4(CO)_9(DMA)]^-$ cluster anion. (Reproduced with permission from ref. 29.)

complexes, this distance is not a strong argument for the formation of a dihydrogen bond. A stronger argument could be based on a stabilization factor. In fact, IR and NMR spectra have demonstrated that the more stable syn conformer exists in solutions. Moreover, the $^1$H $T_{1,min}$ measurements led to the same H$\cdots$H distance of 2.24 Å. Finally, a DFT computational study of the cluster performed at the B3LYP/DZP level has provided us with a syn conformer stabilization energy of 5 kcal/mol. It is interesting that the calculations resulted in an H$\cdots$H distance of 2.134 Å, in good agreement with x-ray and $^1$H $T_{1,min}$ measurements. Experimentally, the energy of dihydrogen bonding has been measured by comparison of the $\nu$(NH) bands observed in the IR spectra of an Re cluster and a pure DMA. Again, the red $\nu$(NH) shift of approximately 100 cm$^{-1}$ corresponds to a $-\Delta H^\circ$ value of 2.2 kcal/mol via the relationship (4.4).

## 5.6. CONNECTION BETWEEN INTRAMOLECULAR DIHYDROGEN BONDING AND DEHYDROGENATION REACTIONS

In previous sections we have shown clearly that intramolecular dihydrogen bonds X−H$\cdots$H−Y, with X and Y representing various chemical elements, can exist in both the solid state and in solution. In addition, the bonds can be a critical factor in the control of molecular conformational states or effects on rapid and reversible hydride−proton exchanges related to the process shown in Scheme 5.1, or the well-known H-D isotope exchanges in similar subsystems [23]. Such bonds could also play an important role in the stabilization of transition states, appearing as a reaction coordinate in many transformations. This is particularly important for systems that easily lose $H_2$ [30]. For example, the Al complex in Structure 5.7, stabilized by proton–hydride interaction, undergoes relatively energetically easy dehydrogenation.

This important aspect has been investigated by Kulkarni [31] on the basis of high-level ab initio calculations performed for complexes $YH_n$–$XH_m$ with Y = Li, B, and Al and X = F, Cl, O, N, and P. The equilibrium structures of the complexes have been optimized at the MP2/6-31++G(d,p) and MP2/6-311++G(2d,2p) levels. Then, beginning with the equilibrium structures, transition states in the dehydrogenation reaction

$$YH_n\text{–}XH_m \rightarrow YH_{n-1}\text{–}XH_{m-1} + H_2 \tag{5.1}$$

have been calculated and their nature has been confirmed by vibrational frequency analysis. Correspondence of the transition states to dehydrogenation has been verified by the intrinsic reaction-coordinate calculations, and the bonding features of transition states have been recognized in terms of a topological analysis of the electron density.

Among the complexes investigated, only three equilibrium structures of the systems $LiH$–$H_2O$, $AlH_3$–$HF$, and $AlH_3$–$HCl$ have actually shown the presence of dihydrogen bonds. It is worth mentioning that these complexes undergo dehydrogenation with smaller activation energy barriers of 0.48, 2.72, and 6.83 kcal/mol, respectively, compared to related compounds without dihydrogen bonding. Two complexes, $LiH$–$H_2O$ and $LiH$–$H_2S$, are the simplest and most representative.

Figures 5.20 illustrates the equilibrium and transition-state structures obtained for these complexes. As shown, the $LiH$–$H_2O$ complex in equilibrium state **1** shows an intramolecular dihydrogen bond with a very short H$\cdots$H distance of 1.580 Å calculated at the MP2/6-311++G(2d,2p) level. The topological analysis of the electron density on the H$\cdots$H direction has resulted in the small $\rho_C$ and positive $\nabla^2 \rho_C$ values (0.0388 and 0.0453 au, respectively) typical of dihydrogen bonding. In contrast, no dihydrogen bonding was observed in the $LiH$–$H_2S$ molecule (**3**), where the corresponding hydrogen atoms are strongly remote.

The transition-state structure of the $LiH$–$H_2O$ complex (**2**), is characterized by a stronger H$\cdots$H interaction where the H$\cdots$H distance is shortened to 1.165 Å. Moreover, in accordance with the idea of intermolecular proton transfer, the O–Li–H angle in the transition state is reduced remarkably (by 7.1°) while the (Li)H$\cdots$H–O angle increases by 12.5°. It is important that topological analysis of the electron density in the transition state reveals a more covalent character for the dihydrogen bond because the $\nabla^2 \rho_C$ value becomes slightly *negative* and the $\rho_C$ parameter increases to 0.0901 au. As follows from the data obtained, even in the transition state of complex $LiH$–$H_2S$, the dihydrogen bond remains very long (2.276 Å) and shows very small $\rho_C$ (0.0139 au) and $\nabla^2 \rho_C$ (0.0249 au) values. All these features are in good agreement with the activation energy of the dehydrogenation, which increases remarkably, from 0.48 kcal/mol in the $LiH$–$H_2O$ system to 2.12 kcal/mol in the $LiH$–$H_2S$ system.

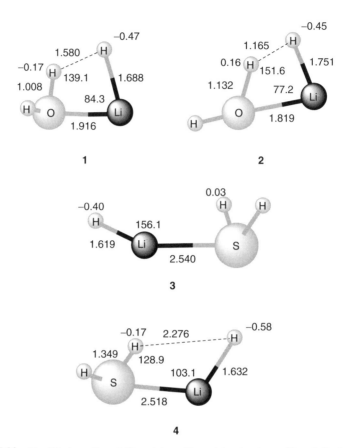

**Figure 5.20** Equilibrium (1 and 3) and transition-state structures (2 and 4) obtained for complexes LiH–H$_2$O and LiH–H$_2$S in the framework of MP2/6-311++G(2d,2p) calculations.

More dramatic changes in activation energies have been found for systems that are not capable of dihydrogen bonding in an equilibrium state. For example, the systems AlH$_3$–PH$_3$ and BH$_3$NH$_3$ have activation energies of 40.81 and 42.5 kcal/mol and the proton transfer becomes endothermic. Thus, intramolecular dihydrogen bonding is actually a *driving force* for the dehydrogenation reaction.

The stabilization of transition states by intramolecular dihydrogen bonding has also been suggested to rationalize the high fluxionality of polyhydride transition metal complexes undergoing fast hydride–hydride exchanges. These exchanges often occur via the *turnstile mechanism*, shown in Scheme 5.2 for a hydride with L = pyridine ligands [32]. Ab initio DFT/B3LYP calculations performed for the Re complexes with R = H and L = 2-aminopyridine and 4-aminopyridine have provided descriptions of the transition-state structures. Figure 5.21 shows the

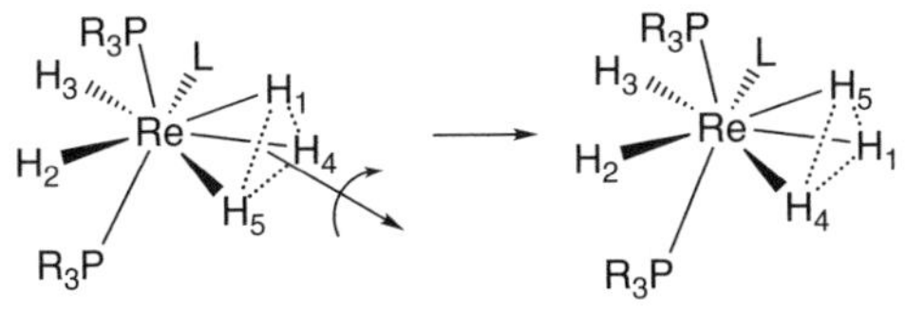

**Scheme 5.2**

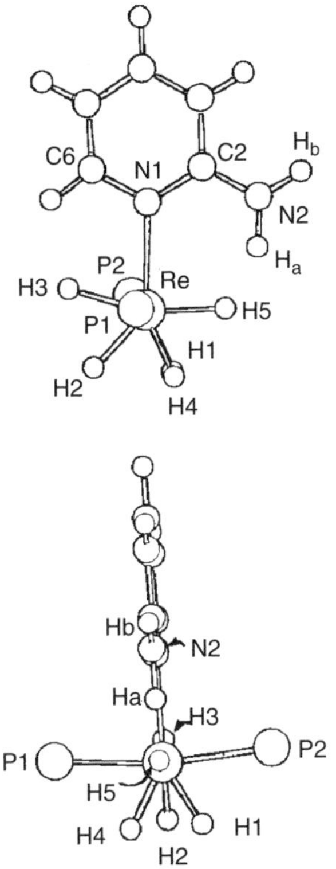

**7ts**

**Figure 5.21** Transition state optimized by ab initio DFT/B3LYP calculations for hydride–hydride exchange via the turnstile mechanism. (Reproduced with permission from ref. 32.)

transition state **7ts** optimized for L = 2-aminopyridine, where the H(5)···H(a) distance is determined to be as short as 1.487 Å. Additionally, formation of this bond, stabilizing the transition state and accelerating hydride–hydride exchange, is well supported by comparison of the bond lengths Re–H(1) (1.652 Å), Re–H(2) (1.652 Å), Re–H(3) (1.705 Å), Re–H(4) (1.652 Å), and Re–H(5) (1.717 Å) on the one hand, and N(2)–H(b) (1.038 Å) and N(2)–H(a) (1.047 Å) on the other.

## 5.7. CONCLUDING REMARKS

1. Strong evidence for the existence of intramolecular dihydrogen interactions C–H$\cdots$H–C, O–H$\cdots$H–C, C–H$\cdots$H–B, N–H$\cdots$H–B, O–H$\cdots$H–B, and X–H$\cdots$H–M with X $=$ O, N and M $=$ metal atom has been obtained experimentally and theoretically for different classes of organic and organometallic compounds. In all cases, these interactions corresponded to the experimental and theoretical criteria, leading to their formulation as dihydrogen bonds.

2. The geometry of intramolecular dihydrogen bonds is governed by the molecular geometry.

3. The bonding energy of these dihydrogen bonds is sufficient to control molecular conformational states.

4. Intramolecular dihydrogen bonds play an important role in molecular systems that lose $H_2$ readily, serving as organizing interactions and factors, and stabilizing transition states in these reactions.

5. Stabilization of transition states by intramolecular dihydrogen bonding explains the high degree of fluxionality of polyhydride transition metal complexes.

6. The relationships observed between interaction energies and H$\cdots$H distances are linear.

## REFERENCES

1. G. J. Kubas, *Metal Dihydrogen and $\sigma$–Bond Complexes*, Kluwer Academic/Plenum Press, New York (2001).
2. D. W. Smith, *J. Chem. Soc. Faraday Trans.* (1996), **92**, 1141.
3. K. N. Robertson, O. Knop, T. S. Cameron, *Can. J. Chem.* (2003), **81**, 727.
4. R. Custelcean, J. E. Jackson, *Chem. Rev.* (2001), **101**, 1963.
5. S. J. Grabowski, A. Pfitzner, M. Zabel, A. T. Dubis, M. Palusiak, *J. Phys. Chem. B* (2004), **108**, 1831.
6. A. H. Pakiari, Z. Jamshidi, *J. Mol. Struct. (Theochem)* (2004), **685**, 155.
7. M. Palusiak, S. J. Grabowski, *J. Mol. Struct. (Theochem)* (2004), **674**, 147.
8. J. G. Planas, C. Vinas, F. Teixidor, M. E. Light, M. B. Hursthouse, *J. Organomet. Chem.* (2006), **691**, 3472.
9. I. R. Padilla-Martinez, M. Rosalez-Hoz, H. Tlahuext, C. Camacho-Camacho, A. Ariza-Castolo, R. Contreras, *Chem. Ber.* (1996), **129**, 441.
10. A. Flores-Para, S. A. Sanchez-Ruiz, C. Guadarrama, H. Noth, R. Contreras, *Eur. J. Inorg. Chem.* (1999), 2069.
11. M. Gilizado-Rodriguez, A. Floes-Parra, S. A. Sanchez-Rutz, R. Tapia-Benavides, R. Contreras, V. I. Bakhmutov, *Inorg. Chem.* (2001), **40**, 3243.
12. M. Guizado-Rodriguez, A. Ariza-Castolo, G. Merino, A. Vela, H. Noth, V. I. Bakhmutov, R. Contreras, *J. Am. Chem. Soc.* (2001), **123**, 9144.
13. J. Li, S. M. Kathmann, G. K. Schenter, M. Gutowski, *J. Phys. Chem. C* (2007), **111**, 3294.

14. P. Singh, M. Zottola, S. Huang, B. R. Show, L. G. Pederson, *Acta Crystallogr. C* (1996), **52**, 693.

15. S. J. Grabowski, *Chem. Phys. Lett.* (2000), **327**, 203.

16. S. Wojtulewski, S. J. Grabowski, *J. Mol. Struct.* (2003), **645**, 287.

17. J. L. Atwood, G. A. Koutsantonis, F. C. Lee, C. L. Ratson, *Chem. Commun.* (1994), 91.

18. J. P. Cambell, J. W. Hwang, V. G. Yuong, R. B. Von Dreele, C. J. Cramer, W. L. Gladfelter, *J. Am. Chem. Soc.* (1998), **120**, 521.

19. R. C. Stevens, R. Bau, R. Milstein, O. Blum, T. F. Koetzle, *J. Chem. Soc. Dalton Trans.* (1990), 1429.

20. R. H. Crabtree, O. Eisenstein, G. Sini, E. Peris, *J. Organomet. Chem.* (1998), **567**, 7.

21. A. J. Lough, S. Park, R. Ramachandran, R. H. Morris, *J. Am. Chem. Soc.* (1994), **116**, 8356.

22. S. Park, A. J. Lough, R. H. Morris, *Inorg. Chem.* (1996), **35**, 3001.

23. W. Xu, A. J. Lough, R. H. Morris, *Inorg. Chem.* (1996), **35**, 1549.

24. S. Park, R. Ramachandran, A. J. Lough, R. H. Morris, *J. Chem. Soc. Chem. Commun.* (1994), 2201.

25. M. J. Calhorda, P. E. M. Lopes, *J. Organomet. Chem.* (2000), **609**, 53.

26. H. S. Chu, C. P. Lau, K. Y. Wong, W. T. Wong, *Organometallics* (1998), **17**, 2768.

27. F. Shi, *Organometallics* (2006), **25**, 4034.

28. S. Aime, E. Diana, R. Gobetto, M. Milanesto, E. Valls, D. Viterbo, *Organometallics* (2002), **21**, 50.

29. M. Panigati, P. Mercandelli, G. D'Alfonso, T. Beringhelli, A. Sironi, *J. Organomet. Chem.* (2005), **690**, 2044.

30. R. H. Crabtree, *Science* (1998), **282**, 2000.

31. S. A. Kulkarni, *J. Phys. Chem. A* (1999), **103**, 9330.

32. R. Bosque, F. Maseras, O. Eisenstein, B. P. Patel, W. Yao, R. H. Crabtree, *Inorg. Chem.* (1997), **36**, 5505.

# 6

# INTERMOLECULAR DIHYDROGEN-BONDED COMPLEXES: FROM GROUPS 1A–4A TO XENON DIHYDROGEN-BONDED COMPLEXES

The role of intermolecular classical hydrogen bonding in inorganic, organic, organometallic, and bioorganic chemistry is well known, whereas intermolecular dihydrogen-bonded systems only begin their development this way. This development is impossible in the absence of comprehensive and detailed representations about the nature of intermolecular dihydrogen bonds and of their geometry, energy, and dynamics. These aspects are the focus of this and subsequent chapters in which we discuss theoretical and experimental data obtained for hydrides of a number of chemical elements capable of accepting protons.

## 6.1. GROUP 1A: DIHYDROGEN BONDS X–H···H–Li AND X–H···H–Na (X = F, Cl, NH₃, CN, NC, HO, HS, ClCC, FCC, HCC)

Hydrogen atoms in the hydrides of alkali metals have the highest negative charges and therefore can be protonated easily even with the weakest proton donors. However, even in this case, the reactivity of these hydrides is too high for experimental studies of dihydrogen bonds that could appear on a reaction coordinate of proton transfer. On the other hand, for example, systems X–H·H–Li or X–H·H–Na are convenient models for theoretical studies, due to the relative simplicity of atoms involved in high-level ab initio calculations. For this reason, the structural properties of dihydrogen bonds X–H···H–Li or X–H···H–Na and their bonding energies are known from the theory, where proton donors HX are often exotic in character from a chemical point of view. It should also be noted that geometry and energy parameters obtained for these bonds

*Dihydrogen Bonds: Principles, Experiments, and Applications*, By Vladimir I. Bakhmutov
Copyright © 2008 John Wiley & Sons, Inc.

depend on the theoretical approach taken, which often affects the final conclusions.

The first study of dihydrogen bonding between LiH and the strong proton donor HF has been carried out by Liu and Hoffman [1]. The authors have performed RHF and MP2 calculations where dihydrogen bonds have not been localized because of the high exothermicity of molecular hydrogen elimination:

$$Li–H + HF \rightarrow LiF + H_2 \qquad (6.1)$$

In contrast, the HF/6-31G** calculations by Rozas and co-workers have resulted in the dihydrogen-bonded complex L–H$\cdots$H–F as a stationary structure [2]. However, when even a weak electric field is applied to this complex, the system changes to L$^+\cdots$H–H$\cdots$F$^-$ with an H$\cdots$H distance of 0.734 Å, very close to that of molecular hydrogen.

Similarly, dihydrogen-bonded complexes of LiH with a variety of proton donors (e.g., HF, HCl, $H_2O$, $H_2S$, and $NH_3$) have been studied by Kulkarni [3] and Kulkarni and Srivastava [4]. Some details of these studies are very interesting and show a dramatic dependence of dihydrogen bonding on the nature of proton donors and the level of theory. All the possible structures for these systems have been optimized at the HF/6-31++G(d,p) and MP2/6-31++G(d,p) levels, and the nature of stationary points has been examined by calculating their vibrational frequencies at the MP2/6-31++G(d,p) level.

The systems Li–H·HF and Li–H·$H_2O$ are particularly representative. In accordance with the Mulliken charges obtained for the isolated molecules

$$^{(+0.54)}L\text{–}H^{(-0.54)} \qquad ^{(+0.41)}H\text{–}F^{(-0.41)} \qquad ^{(+0.36)}H\text{–}O^{(-0.72)}\text{–}H^{(+0.36)} \qquad (6.2)$$

demonstrating their high ability for dihydrogen bonding, the linear complex L–H$\cdots$H–F has actually been localized as a stationary point in the HF calculations. In contrast, none was found by using the MP2 calculations. Thus, the dihydrogen-bonded structure Li–H$\cdots$H–F is an *artifact* of the HF theory; in other words, the inclusion of *electron correlation is vital for validation of the dihydrogen bonding*.

MP2/6-31++G(d,p) treatments of the system Li–H·H–Cl have resulted in the dihydrogen-bonded adduct shown in Structure 6.1, indicating an almost linear geometry with a stabilization energy of −8.98 kcal/mol. However, the complex has shown one imaginary frequency, corresponding to its transition-state nature. In other words, it does not contribute to the ground-state structure of the Li–H·HCl system [4].

1.328 Å  1.413 Å  1.618 Å

Cl ——— H $\cdots$ H ——— Li

**Structure 6.1**

**Structure 6.2**

Three stationary points have been found on the potential energy surface of the system LiH·H$_2$O [3]. They did not show dihydrogen bonding, except as shown in Structure 6.2. The latter could represent a bifurcated dihydrogen-bonded complex with very long H···H distances (2.325 Å) and a formation energy of −5.51 kcal/mol. However, this structure shows two imaginary frequencies and hence should be ruled out in the context of a ground state.

An interesting product of MP2 optimizations, located at an energy minimum on the potential energy surface of the system H−Li·H$_2$O, is the complex containing the Li−O bond:

$$^{(-0.71)}H-Li^{(+0.61)}\_^{(-0.85)}OH(H)^{(+0.47)} \tag{6.3}$$

The complex is characterized by a large energy gain, −18.2 kcal/mol, and its hydride atom shows a larger negative charge then that found in the isolated Li−H molecule [see (6.2)], thus could be a good candidate for dihydrogen bonding. In accordance with the expectation, the dimer (Li−H···H$_2$O)$_2$, formed in a head-to-tail manner and located at minimum energy on the potential energy surface, is actually a dihydrogen-bonded complex with the cyclic $C_{2h}$ eight-membered structure shown in Figure 6.1, where the H···H contacts are very short (1.192 Å). The fragment (Li)H···H−O is practically linear, with the (Li)H···H−O angle equal to 176.8° as in the case of classical hydrogen bonds, whereas the angle (O)H···H−Li is strongly bent (113.3°). Due to double dihydrogen bonding, the structure is stabilized by an energy of −45.34 and −40.01 kcal/mol at the MP2 and CCD(T) levels, respectively. Such a high stabilization energy is typical of the strongest ionic hydrogen bonds (e.g., F···H···F$^-$), which cannot

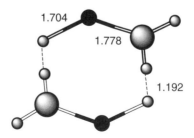

**Figure 6.1** Stationary structure of dimer (LiH−H$_2$O)$_2$ with short H···H distances obtained by calculations at the MP2/6-31++G(d,p) level. The black spheres represent the Li atoms and the largest spheres are the O atoms.

be distinguished from covalent chemical bonds. In accord with this statement, a topological analysis of the electronic density in the dihydrogen-bonded dimer $(LiH-H_2O)_2$ also led to unusual results. In fact, the bond critical points, obtained in the $H\cdots H$ directions at the MP2/6-31++G(d,p) level, are characterized by a very small electron density with a $\rho_C$ value of 0.0756 au versus the 0.046 au obtained, for example, in the dihydrogen-bonded system $Li-H\cdots HNH_3^+$, which has the longer $H\cdots H$ distance of 1.390 Å [5]. In addition, the Laplacian in the $H\cdots H$ directions of $(LiH-H_2O)_2$ has been found to be *negative* ($\nabla^2\rho_C$ value $= -0.0252$ au), which is typical for a covalent bond.

The result of an energy decomposition analysis performed for this complex was also unusual. In contrast to numerous dihydrogen-bonded systems with significant predominance of the electrostatic interaction, the dihydrogen-bonded dimer $(LiH\cdot H_2O)_2$ has shown the charge transfer contribution to exceed the electrostatic energy: $-125.30$ versus $-81.40$ kcal/mol, respectively.

The system $H_2S\cdots LiH$ [4], which contains the weak proton donor $H_2S$, has also shown three stationary geometries on the potential energy surface. However, in this case, two of them were the dihydrogen-bonded complexes depicted in Structures 6.3 and 6.4. Structure 6.4 represents a monodentate complex with an $H\cdots H$ distance of 1.918 Å. Its bonding energy is calculated as $-3.7$ kcal/mol. The second complex with the longer $H\cdots H$ distance of 2.674 Å is bidentate and its energy is smaller ($-2.31$ kcal/mol). However, again, both complexes do not contribute to the structure of the $H_2S\cdot LiH$ system because they show imaginary frequencies. Finally, a dihydrogen-bonded dimer similar to $(LiH\cdot H_2O)_2$ has not been found in calculations of the $H_2S\cdot LiH$ system.

Grabowski has performed a comprehensive theoretical study of dihydrogen-bonded complexes formed by LiH or NaH with an HF molecule as a proton donor, to clarify the influence of the theory level [6]. Figure 6.2 summarizes the data obtained for different dihydrogen-bonded complexes, the geometries of which have been fully optimized. First, there are no significant differences in the geometry of the donating (HF) and accepting (NaH or LiH) molecules obtained by various methods but in the same basis set. Only the MP2/6-31G** level shows slightly different results, demonstrating that the diffuse functions play an important role in descriptions of such systems. By contrast, the $H\cdots H$ bond length

**Structure 6.3**

**Structure 6.4**

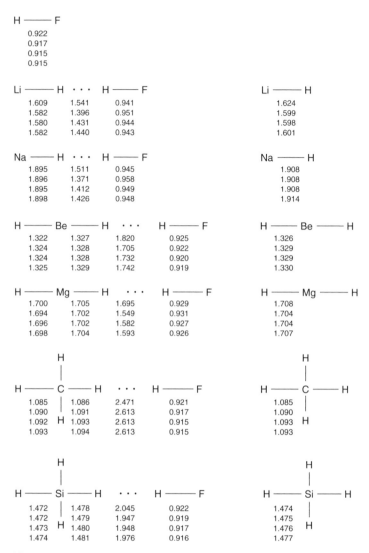

**Figure 6.2** Geometries of isolated molecules and dihydrogen-bonded complexes optimized by calculations at (from top to bottom) the MP2/6-31G**, MP2/6-311++G**, MP4(SDQ)/6-311++G**, and QCISD/6-311++G** levels. (Reproduced with permission from ref. 6.)

in dihydrogen-bonded systems changes more notably. Nevertheless, as follows from Figure 6.2, all the theory levels lead to a practically linear geometry of the dihydrogen bonds, where a smaller H···H distance corresponds to a longer H–F bond. This is typical of classical hydrogen bonds. It is interesting that the H–F bond elongation in Li–H···H–F and Na–H···H–F complexes is quantitatively close to that calculated for the strongest hydrogen-bonded systems, such as $(F \cdots H \cdots F)^-$ and $(Cl \cdots H \cdots F)^-$. Alkorta and co-workers [7] have shown that the effect of H–X bond elongation accompanied by red-shifted X–H stretching frequencies is common and observed even for weak C–H acids. For example, compared to an isolated HCCH molecule, the C–H distance increases by 0.012 Å in complex HCCH$\cdots$H–Li.

The situation is more complicated for proton-accepting components. As shown in Figure 6.2, the formation of neutral dihydrogen-bonded complexes leads to shortening bonds Na–H and Li–H. In contrast, the charged systems LiNCH$^+\cdots$H–Li and NaNCH$^+\cdots$H–Li [7] show the Li–H bonds to be elongated by 0.023 and 0.017 Å, respectively. Thus, changes in the M–H bond length of the proton acceptor do not follow a simple pattern, although the corresponding stretching frequencies appear systematically blue-shifted.

A topological analysis of the electron density in the framework of AIM theory, performed for the systems in Figure 6.2, has completely confirmed their formulation as dihydrogen-bonded complexes. In accord with the AIM criteria, the $\rho_C$ and $\nabla^2 \rho_C$ parameters at the bond critical points found in the H$\cdots$H directions are typical of dihydrogen bonds: 0.042 and 0.057 au for complex LiH$\cdots$HF and 0.046 and 0.048 au for complex NaH$\cdots$HF, respectively. The presence of the bond critical points can be well illustrated by the molecular graph in Figure 6.3, obtained for the HCCH$\cdots$H–Li complex by Grabowski and co-workers [8].

For comparison, the authors have probed a complex formed by the same proton-donor molecule and molecular hydrogen. In this very weak complex, HCCH$\cdots$(H$_2$), the H$\cdots$(H$_2$) distance has been calculated as 2.606 Å (i.e., significantly larger than the sum of the van der Waals radii of H). It is extremely interesting that a topological analysis of the electron density also leads to the appearance of the bond critical point in the H$\cdots$(H$_2$) direction. However, the $\rho_C$ and $\nabla^2 \rho_C$ values are very small (0.0033 and 0.0115 au, respectively) compared with those in the HCCH$\cdots$HLi complex (0.0112 and 0.0254 au, respectively). The most important conclusion of this comparison is: *There is no evident borderline between the dihydrogen-bonded complexes and the van der Waals systems.*

The effects, caused by methods of calculations, on energies of dihydrogen bonding in complexes of Li and Na corrected for BSSE are shown in Table 6.1. It is clearly seen that only the MP2/6-31G** level gives remarkable deviations. On the other hand, it has been found that the MP2/6-311++G** method is sufficient to describe dihydrogen-bonded complexes.

In addition to monomeric linear dihydrogen-bonded complexes **I** shown in Figure 6.4, Alkorta and co-workers [9] have investigated associated systems **II** and **III**, where HX and HY are proton acceptors (LiH and NaH) and proton donors (see Table 6.2), respectively. According to the MP2/6-311++G(2 d,2p)

**Figure 6.3**   Molecular graph of the HCCH$\cdots$H–Li complex. The large spheres are atoms and the small spheres are bond critical points.

**TABLE 6.1. Total Energies of Dihydrogen-Bonded Complexes Formed Between LiH and NaH Hydrides and the HF Molecule as a Function of Levels of Theory**

| Complex/Method | BSSE (kcal/mol) | Total Energy (kcal/mol) |
|---|---|---|
| LiH···HF | | |
| MP2/6-31G* * | 0.88 | −10.12 |
| MP2/6-311++G* * | 0.72 | −12.62 |
| MP4/6-311++G* * | 0.72 | −12.47 |
| QCISD/6-311++G* * | 0.71 | −11.61 |
| QCISD(T)/6-311++G* * | 0.73 | −11.94 |
| NaH···HF | | |
| MP2/6-31G* * | 0.88 | −10.82 |
| MP2/6-311++G* * | 0.76 | −13.81 |
| MP4/6-311++G* * | 0.74 | −12.74 |
| QCISD/6-311++G* * | 0.70 | −12.28 |
| QCISD(T)/6-311++G* * | 0.72 | −12.66 |

*Source:* Ref. 6.

X—H ··· H—Y

I

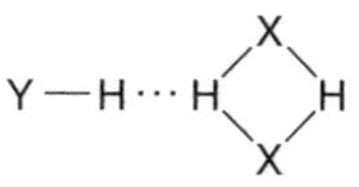

II

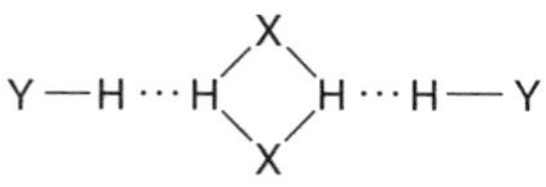

III

**Figure 6.4**

calculations, the systems occupy minimum energies on the potential energy surface and show geometries with symmetry $C_{\infty v}$ (I), $C_{2v}$ (II), and $D_{2h}$ (III). The formulation of the systems as dihydrogen-bonded complexes has been confirmed by independent topological analysis of the electron density.

Table 6.2 shows the H···H distances and formation energies obtained for the monomer and more-associated systems. The shortest H···H contacts, 1.5 to 1.6 Å, have been calculated for all of the monomer and associated dihydrogen-bonded systems formed by the strongest proton donor, HNC, followed by the weaker

**TABLE 6.2. H···H Distances and Interaction Energies Obtained for Structures I to III in Figure 6.4**

| | | Distance (Å) end Energy (kcal/mol) for Structure: | | |
|---|---|---|---|---|
| X–H | Y–H | I | II | III |
| LiH | HCN | 1.790 (−8.25) | 1.870 (−6.08) | 1.894 (−11.27) |
| LiH | HNC | 1.487 (−13.32) | 1.538 (−10.72) | 1.573 (−19.62) |
| LiH | HCCH | 1.995 (−3.97) | 2.070 (−3.18) | 2.080 (−6.22) |
| NaH | HCN | 1.770 (−9.02) | 1.812 (−7.85) | 1.836 (−14.74) |
| NaH | HNC | 1.451 (−14.71) | 1.456 (−13.78) | 1.497 (−25.7) |
| NaH | HCCH | 1.987 (−4.26) | 2.033 (−3.85) | 2.038 (−7.80) |

[a] At the MP2/6-311++G(2 d,2p) level.

donor, HCN. At the same time, the longest distances are observed for the weakest proton-donor molecule, HCCH. It is remarkable that the H···H distances increase from **I** to **III** while the energy gain decreases from **I** to **II** but increases again in **III**. These results show clearly important contributions coming from the molecular electrostatic potential of the isolated proton-donor molecule.

Table 6.3 concludes this section and contains geometrical and energetic parameters obtained for various dihydrogen-bonded complexes of LiH and NaH at the MP2/6-311++G** level. It is worth mentioning that the energies in Table 6.3 have been obtained as differences between the total energies of the complexes and the energies of isolated monomers.

Despite the above-mentioned dependence of the results on the computing method and the fact that some of these systems have not been localized at minima of the potential energy surface, the data in Table 6.3 illustrate well the structural and energetic tendencies in dihydrogen bonding with the participation of Li and Na hydrides. As shown, the dihydrogen bonds change from medium to very strong as a function of the strength of proton donors. Compare, for example, the interaction energy from number 24 to numbers 22, 21, 19, and 16. The dihydrogen bonds show very short H···H contacts up to 1.309 Å in the $LiNCH^+$···H–Li system. Even in the case of very weak proton donors such as HCCH or $H_3SiCCH$, the H···H distance is remarkably smaller than the sum of the van der Waals radii of H. Finally, the interaction energy increases slightly and the H···H distance reduces on going from Li to Na (see, e.g., numbers from 4 to 2, or from 11 to 7 and from 16 to 14).

## 6.2. GROUP 2A: DIHYDROGEN BONDS X–H···H–M AND X–H···H–B (X = F, Cl, Br, NH₃, NNN, CN, NC, ClCC, FCC, HCC, CH₃CC, F₂Be, FKr, FAr)

As in the case of alkali metals, hydrogen atoms attached to the elements of group 2A have pronounced negative charges. They are, however, smaller than

**TABLE 6.3. BSSE-Corrected Interaction Energies and H···H Distances for Li and Na Dihydrogen-Bonded Complexes Formed with Various Proton-Donor Molecules**[a]

| Number | | $\Delta E$ (kcal/mol) | r(H·H)Å | Ref |
|---|---|---|---|---|
| 1 | LiNCH+···H–Li | 27.1[b] | 1.309 | 7 |
| 2 | F–H···H–Na | 10.82 | 1.371 | 10 |
| 3 | +NH3–H···H–Li | 38.08 | 1.390 | 5 |
| 4 | F–H···H–Li | 10.12 | 1.396 | 10 |
| 5 | Cl–H···HLi | 9.05 | 1.413 | 3 |
| 6 | CNH···H–Na | 14.7 | 1.451 | 9 |
| 7 | CNH···H–Na | 15.8[b] | 1.426 | 7 |
| 8 | HOH···H–Na | 15.7[b] | 1.428 | 7 |
| 9 | NaNCH+···H–Li | 23.7[b] | 1.429 | 7 |
| 10 | HOH···H–Li | 18.6[b] | 1.454 | 7 |
| 11 | CNH···H–Li | 14.2[b] | 1.468 | 7 |
| 12 | CNH···H–Li | 13.3 | 1.487 | 9 |
| 13 | (LiH · H2O)2[c] | 20.77[d] | 1.704 | 3 |
| 14 | NCH···H–Na | 9.7[b] | 1.754 | 7 |
| 15 | NCH···H–Na | 8.06 | 1.770 | 9 |
| 16 | NCH···H–Li | 8.8[b] | 1.774 | 7 |
| 17 | NCH···H–Li | 7.42 | 1.970 | 5 |
| 18 | NCH···H–Li | 8.2 | 1.790 | 9 |
| 19 | NCCCH···H–Li | 8.1[b] | 1.809 | 7 |
| 20 | HS–H···H–Li | 2.93 | 1.918 | 4 |
| 21 | ClCCH···H–Li | 5.0[b] | 1.930 | 7 |
| 22 | FCCH···H–Li | 4.8[b] | 1.950 | 7 |
| 23 | FCCH···H–Li | 3.97 | 2.0165 | 8 |
| 24 | HCCH···H–Li | 4.4[b] | 1.978 | 7 |
| 25 | HCCH···H–Li | 3.14 | 2.333 | 5 |
| 26 | HCCH···H–Li | 3.97 | 1.995 | 9 |
| 27 | HCCH···H–Li | 3.65 | 2.0499 | 8 |
| 28 | HCCH···H–Na | 4.26 | 1.987 | 9 |
| 29 | H3SiCCH···H–Li | — | 2.134 | 11 |

[a]Obtained at the MP2/6-311++G* * level.
[b]Calculated at the MP2/aug-cc-pVTZ level.
[c]See Figure 6.1.
[d]Accounting for two H···H contacts.

in alkali metal hydrides. For example, in the framework of MP2/6-31** treatment, Robertson and co-workers [11] have obtained the corresponding Mulliken charge in dihydride molecule H–Be–H as −0.093e versus −0.171 e in the Li–H hydride. Alkorta and co-workers [5] reported slightly different results but with the same tendency: The negative charge decreases from Li–H (−0.193 e) to H–Be–H (−0.109 e). This simple consideration shows that the H–Mg–H and H–Be–H hydrides can accept a proton to yield dihydrogen bonds, but they will obviously be weaker than those formed by metal hydrides of group 1A.

In the first report on the beryllium dihydrogen-bonded complex, published in 1996 [5], the $H-Be-H\cdots H-N(H_3)^+$ system was located at an energy minimum of the potential energy surface calculated at the MP2/6-311++G** level. By analogy with classical hydrogen bonds, this dihydrogen complex can be said to be charge-assisted.

The equilibrium structure of the dihydrogen bond has shown a linear geometry and the $H\cdots H$ distance and the bonding energy have been calculated as 1.744 Å and $-7.88$ kcal/mol, respectively. The formulation of this complex as a dihydrogen-bonded system has been confirmed by topological analysis of the electron density, where the $\rho_C$ value was obtained as 0.025 au and the Laplacian, $\nabla^2\rho_C$, was positive. The plot of the Laplacian, $\nabla^2\rho_C$, is shown in Figure 6.5, where the dihydrogen bond is shown clearly.

Recently, Hayashi and co-workers reached a more comprehensive description of the same complex [12] using ab initio molecular dynamics (MD) simulation at a finite temperature (300 K). This study was performed to reveal the influence of a vibrational contribution, which could not be negligible in the case of dihydrogen-bonded systems. Since these systems generally have low-frequency modes, they act as floppy molecules involving wide-amplitude vibration. Additionally, the authors have employed ab initio path integral molecular dynamics (PIMD) simulation for the quantitative analysis of molecular vibrations due to thermal and quantum fluctuations of nuclei within the framework of Born–Oppenheimer approximation. It should be emphasized that these approaches have independently confirmed the existence of the dihydrogen-bonded complex $NH_4^+\cdots H-Be-H$. The electronic structure of the complex has been obtained by calculations at the MP2/6-311++G** level, and the interaction energy has been determined as $-9.6$ kcal/mol, which is slightly larger than in the data of Alkorta et al.

Figure 6.6 illustrates the $NH_4^+\cdots H-Be-H$ dihydrogen complex described in the framework of regular ab initio calculations and PIMD theory. Table 6.4

**Figure 6.5** Plot of the Laplacian, $\nabla^2\rho_C$, obtained for the dihydrogen complex $H-Be-H\cdots H-N^+(H_3)$ at the MP2/6-311++G** level. (Reproduced with permission from ref. 5.)

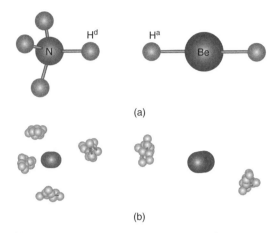

(a)

(b)

**Figure 6.6** (a) Dihydrogen-bonded complex system $NH_4^+ \cdots H-Be-H$ represented by the optimized equilibrium geometry, and (b) the ab initio path integral molecular dynamics simulation (representative snapshot of configuration). (Reproduced with permission from ref. 12.)

**TABLE 6.4. Geometry Parameters of the Dihydrogen-Bonded Complex $[NH_3H^d]^+ \cdots {}^aH-Be-H$ Optimized for the Equilibrium and Dynamic Structures in Figure 6.6**

| Parameter | Equilibrium | Classical | Quantum |
|---|---|---|---|
| $r(H \cdots H)$ (Å) | 1.571 | 1.798 | 1.790 |
| $\Delta r(H \cdots H)^a$ (Å) | — | 0.252 | 0.291 |
| $r(N-Be)$ (Å) | 3.964 | 3.955 | 3.968 |
| $\Delta r(N-Be)$ (Å) | — | 0.179 | 0.238 |
| $N-Be-H^a$ (deg) | 0.0 | 16.7 | 18.8 |
| $\Delta(N-Be-H^a)^a$ (deg) | — | 8.7 | 9.2 |
| $H^d-N-Be$ (deg) | 0.0 | 20.4 | 20.4 |
| $\Delta(H^d-N-Be)$ (deg) | — | 12.3 | 12.4 |

$^a \Delta$ represents the dispersion of the bond lengths and angles.

lists the geometrical parameters of the corresponding equilibrium and dynamic structures. As shown, the $H \cdots H$ distance obtained by the MD method is longer by $\sim 0.2$ Å than that obtained for the optimized equilibrium structure, due to thermal effects (classical). In turn, PIMD gives the longest distance because of the zero-point motion under the anharmonic potential. At the same time, the effects on the $Be \cdots N$ distance are not significant. In some sense these effects are related to effective elongations of internuclear (e.g., $H \cdots H$) distances determined by $^1H$ NMR relaxation measurements in solution.

Dispersions of bond lengths are also maximal when using the PIMD method. It is interesting that the bond angles $N-B-H^a$ and $H^dNBe$ are determined by the PIMD and MD methods as $16°$ and $20°$, respectively. In other words, the average structures for the $N-H^d \cdots H^a - Be$ fragment obtained in the framework of the PIMD and MD methods are not linear, in contrast to the linear equilibrium geometry.

The same $BeH_2$ model has been used by Grabowski as a proton acceptor [13] to investigate the effects of variation in the strength of proton-donor sites on dihydrogen bonding. Table 6.5 shows the topological parameters of the electron density, the $H \cdots H$ distance, and the $Be-H$ bond length obtained for equilibrium geometries at the MP2/6-311++G** level. First, the topological parameters $\rho_C$ and $\nabla^2 \rho_C$ are in a good agreement with the formulation of these complexes as dihydrogen bonded. Second, there is a clear tendency to reduce the $H \cdots H$ distance and increasing the electron density in the $H \cdots H$ directions with increasing proton-donor strength. At the same time, Table 6.5 shows that the $Be-H$ bond length is not as sensitive to these effects.

Due the presence of three F atoms, the $NF_3H^+$ molecule is probably one of the strongest proton donors. For this reason, dihydrogen bonding to the $BeH_2$ hydride is surprisingly unusual. According to calculations at the 6-311++G** and aug-cc-pVDZ levels, the $NF_3H^+ \cdots H-BeH$ molecule exhibits an extremely

**TABLE 6.5. Effects of Proton-Donor Strength on Dihydrogen Bonding in BeH$_2$ Complexes**

| Proton Donor | H$\cdots$H (Å) | Be–H (Å) | $\rho_C$ (au) | $\nabla^2 \rho_C$ (au) |
|---|---|---|---|---|
| HBF$_2$ | 2.834 | 1.330 | 0.0024 | 0.0070 |
| HCCH | 2.289 | 1.329 | 0.0056 | 0.0150 |
| HBr | 2.052 | 1.330 | 0.0096 | 0.024 |
| HCN | 2.130 | 1.331 | 0.0076 | 0.0205 |
| HCl | 1.947 | 1.330 | 0.0114 | 0.0280 |
| HNNN | 1.914 | 1.331 | 0.0122 | 0.0316 |
| HF | 1.705 | 1.328 | 0.0164 | 0.0488 |

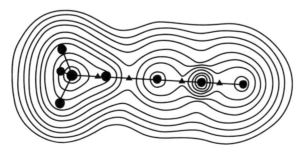

**Figure 6.7** Contour map of the electron density obtained for the $NF_3H^+ \cdots H-Be-H$ dihdyrogen-bonded complex where attractors (nuclei) are shown as circles and bond critical points are shown as triangles. (Reproduced with permission from ref. 14.)

short H$\cdots$H distance of 1.114 and 1.132 Å, respectively [14]. Moreover, the H$\cdots$H bond critical point, shown clearly on the contour map of the electron density in Figure 6.7, is characterized by a relatively high electron density ($\rho_C = 0.0868$ and 0.0821 au at the 6-311++G** and aug-cc-VDZ levels, respectively) and the Laplacian $\nabla^2 \rho_C$ is negative ($-0.0542$ and $-0.0248$ au, respectively). As in the case of the unusual dihydrogen-bonded system (LiH–H$_2$O)$_2$, the Laplacian values correspond to the remarkable covalent character of the charge-assisted dihydrogen bond in NF$_3$H$^+$$\cdots$H–BeH. One can postulate that this system reflects the situation when two hydrogen atoms move toward molecular hydrogen.

Dihydrogen bonding in the more-associated systems, containing the monomer proton acceptor H–X, the dimer (XH)$_2$ (where X = Be, Mg), and one or two proton-donor molecules H–Y, has been studied by Alkorta and co-workers [9]. According to calculations performed at the MP2/6-311++G(2 d,2p) level, the Be and Mg hydrides give structures **I** to **III** in Figure 6.4, which are similar to their Li and Na analogos. In addition, they form complexes **IV** to **VI** in Figure 6.8. It is worth noting that most of complexes occupy true energy minima on the potential energy surface. Only Be dihydrogen-bonded complexes (**III**) have shown two imaginary frequencies, and the Be complexes of structure **VI** do not exist. Comparison of the energy parameters in Tables 6.2 and 6.6 shows that the dihydrogen-bonded complexes of Be and Mg are considerably weaker

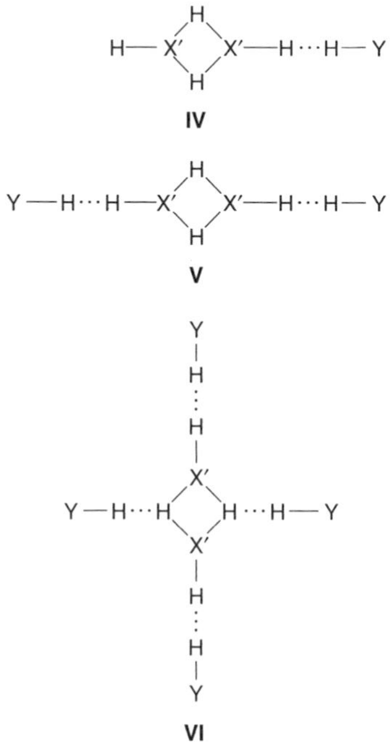

**Figure 6.8**

**TABLE 6.6. Interaction Energies[a] for Be and Mg Structures I to VI in Figures 6.4 and 6.8**

| X–H | Y–H | Interaction Energy (Kcal/mol) in Structure: | | | | | |
|-----|-----|------|------|------|------|------|------|
|     |     | I    | II    | III   | IV    | V     | VI    |
| HBeH | HCN  | −1.89 | −0.05 | 0.36  | −2.42 | −4.40 |       |
| HBeH | HNC  | −3.23 | −0.31 | 0.29  | −4.02 | −7.25 |       |
| HBeH | HCCH | −1.03 | −0.14 | −0.21 | −1.27 | −2.46 |       |
| HMgH | HCN  | −3.66 | −1.1  | −1.46 | −3.99 | −7.58 | −6.24 |
| HMgH | HNC  | −6.24 | −2.68 | −3.90 | −6.72 | −12.7 | −12.6 |
| HMgH | HCCH | −1.89 | −0.77 | −1.46 | −2.03 | −3.99 | −5.19 |

[a]Obtained at the MP2/6-311++G(2 d,2p) level.

than those of their Na and Li relatives. However, their tendency to change in strength with variation in their proton donors is almost the same. As in the case of the Li and Na dihydrogen-bonded systems, changes in the total formation energies are controlled mainly by the molecular electrostatic potential of isolated proton-donor molecules.

It is well known that rare-gas compounds such as HArF or HKrF exhibit the ion-pair character $(HAr)^+F^-$. Therefore, they are of great interest as proton donors in dihydrogen bonding [15]. The equilibrium structures of systems $BeH_2 \cdot HArF$ and $BeH_2 \cdot HKrF$ and of isolated components have been optimized at the MP2/6-311++G(2 d,2p) level. The calculations led to dihydrogen-bonded complexes with short H···H distances: 1.312 and 1.590 Å for H–Be–H···HArF and H–Be–H···HKrF, respectively. The interaction energies have been determined as −5.28 and −1.70 kcal/mol (see Table 6.7).

Figure 6.9 schematically illustrates the equilibrium structures optimized for the FArH···H–BeH dihydrogen-bonded complex and individual molecules $BeH_2$ and HArF. Comparing these structures reveals the following important features, accompanying the complexation: The Ar–F and H–Ar bond lengths increase, and

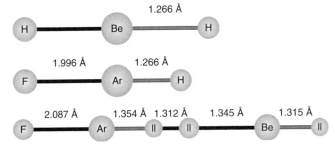

**Figure 6.9** Equilibrium structures optimized at the MP2/6-311++G(2 d,2p) level for the dihydrogen-bonded Ar complex and the isolated molecules $BeH_2$ and HArF. (Reproduced with permission from ref. 15.)

the Be–H bond participating in the dihydrogen bonding also elongates, whereas the free Be–H bond is shortened by 0.01 Å. Similar data have been obtained for FKrH···H–BeH bonding. In addition, the stretching frequencies calculated for all of the bonds predict *red shifts* for ν(H–Ar), ν(H–Kr), ν(Ar–F), and ν(Kr–F), in contrast with the *blue shifts* expected for ν(Be–H).

The very weak complexes FArCCH···H–Be–X (X = H, F, Cl, Br) have also been found through MP2/6-311++G(2 d,2p) calculations [15]. Again, the equilibrium structures of the complexes are linear, and the interaction energy calculated (see Table 6.7) decreases reasonably in the order

$$\text{FArCCH} \cdots \text{H–Be–H} > \text{FArCCH} \cdots \text{H–Be–Br} > \text{FArCCH} \cdots \text{H–Be–Cl} >$$

$$\text{FArCCH} \cdots \text{H–Be–F} \tag{6.4}$$

Table 6.7 summarizes the energetic and geometric parameters obtained at the MP2/6-311++G** level for various dihydrogen-bonded systems of Mg and Be. These data allow us to trace the effects of proton-donor molecules and metal atoms on dihydrogen bonding. As shown in the table, these dihydrogen bonds can be classified from weak to medium. Two of these systems are different: $NF_3H^+$···H–BeH shows a very strong dihydrogen bond with a bonding energy of −20.6 kcal/mol, whereas the system $F_2BH$···H–Be–H is extremely weak, due to the small hydridicity of the hydrogen atoms in $HBF_2$. The effect of two F atoms is obvious.

One unusual characteristic of molecular systems 34 to 36 requires additional explanation. Complexes 34 and 35 show H···H distances that are very close to the sum of the van der Waals radii of H, 2.4 Å whereas the distance in system 36 is notably larger. The formulation of this system as dihydrogen-bonded is particularly questionable because the H···H separation is even longer than the 2.7 Å suggested by Alkorta et al. as the sum of the van der Waals radii of H calculated for the system $H_3C$–H···H–$CH_3$, with an interaction energy of 0.2 kcal/mol [17].

In this context it is interesting to compare the energy contributions to the total bonding energies of three complexes that are formed by the same proton donor, FCCH, and different proton acceptors: Li–H, $BeH_2$, and $H_2$. It is obvious that the FCCH···σ($H_2$) complex is too weak and represents a van der Waals system [8]. It follows from the energy parameters shown in Table 6.8 that the electrostatic term in the LiH dihydrogen complex outweighs the exchange term, which corresponds to steric repulsion between two charge clouds. The situation changes in the $BeH_2$ dihydrogen-bonded complex. Here the ES term is only slightly larger than the EX term. Finally, in the FCCH···σ($H_2$) complex, the exchange term outweighs the ES term. Thus, the nature of a proton acceptor is an important factor controlling the character of dihydrogen bonding: The complexes are stable, due to the electrostatic interactions in the LiH···HCCF system, whereas all the contributions are important for the systems FCCH···H–Be–H and FCCH···σ($H_2$).

**TABLE 6.7. BSSE-Corrected Interaction Energies and H· · ·H Distances Obtained for Mg and Be Dihydrogen-Bonded Complexes Formed with Various Proton-Donor Molecules**

| Number | System | $-\Delta E$ (kcal/mol) | $r(\text{H} \cdots \text{H})$ (Å) | Ref. |
|--------|--------|------------------------|-----------------------------------|------|
| 1 | $NF_3H^+ \cdots H-Be-H$ | 20.6 | 1.114 | 14 |
| 2 | $FArH \cdots H-Be-H$ | 5.28 | 1.312 | 15 |
| 3 | $F-H \cdots H-Mg-H$ | 6.02 | 1.549 | 6 |
| 4 | $NH_4^+ \cdots H-Be-H$ | 9.2 | 1.571 | 14 |
| 5 | $NH_4^+ \cdots H-Be-H$ | 9.26 | 1.591 | 5 |
| 6 | $FKrH \cdots H-Be-H$ | 1.70 | 1.591 | 15 |
| 7 | $NH_4^+ \cdots H-Be-F$ | 6.0 | 1.620 | 14 |
| 8 | $FOH \cdots H-Be-H$ | 4.83 | 1.647 | 13 |
| 9 | $CNH \cdots H-Mg-H$ | 6.24 | 1.652 | 9 |
| 10 | $FH \cdots H-Be-H$ | 2.94 | 1.705 | 13 |
| 11 | $FH \cdots H-Be-H$ | 2.94 | 1.705 | 6 |
| 12 | $CNH \cdots H-Be-H$ | 3.23 | 1.777 | 9 |
| 13 | $CNH \cdots H-Be-H$ | — | 1.864 | 11 |
| 14 | $NNNH \cdots H-Be-H$ | 2.03 | 1.914 | 13 |
| 15 | $NCH \cdots H-Mg-H$ | 3.66 | 1.944 | 9 |
| 16 | $ClH \cdots H-Be-H$ | 1.69 | 1.947 | 13 |
| 17 | $BrH \cdots H-Be-H$ | 1.45 | 2.052 | 13 |
| 18 | $NCH \cdots H-Be-H$ | 1.89 | 2.055 | 9 |
| 19 | $FArCCH \cdots H-Be-H$ | 1.58 | 2.097 | 16 |
| 20 | $FArCCH \cdots H-Be-Cl$ | 1.15 | 2.111 | 16 |
| 21 | $NCH \cdots H-Be-H$ | 1.53 | 2.115 | 5 |
| 22 | $FArCCH \cdots H-Be-F$ | 1.07 | 2.120 | 16 |
| 23 | $FArCCH \cdots H-Be-Br$ | 1.19 | 2.126 | 16 |
| 24 | $HCCH \cdots H-Mg-H$ | 1.82 | 2.126 | 9 |
| 25 | $NCH \cdots H-Be-H$ | 1.68 | 2.130 | 13 |
| 26 | $HCCH \cdots H-Be-H$ | 1.03 | 2.225 | 9 |
| 27 | $FCCH \cdots H-Be-H$ | — | 2.242 | 11 |
| 28 | $H_3SiCCH \cdots H-Be-H$ | — | 2.252 | 11 |
| 29 | $FCCH \cdots H-Be-H$ | 0.91 | 2.293 | 8 |
| 30 | $FCCH \cdots H-Be-F$ | 0.63 | 2.276 | 8 |
| 31 | $HCCH \cdots H-Be-H$ | 0.92 | 2.289 | 13 |
| 32 | $HCCH \cdots H-Be-H$ | 0.91 | 2.293 | 8 |
| 33 | $HCCH \cdots H-Be-F$ | 0.61 | 2.299 | 8 |
| 34 | $CH_3CCH \cdots H-Be-H$ | — | 2.315 | 11 |
| 35 | $LiCCH \cdots H-Be-H$ | — | 2.472 | 11 |
| 36 | $F_2BH \cdots H-Be-H$ | 0.23 | 2.834 | 13 |

**TABLE 6.8. Energy Contributions to the Total Energy Obtained for Three Complexes Using Morokuma's Analysis**

| Energy Contribution[a] | Interaction Energy (kcal/mol) for the Complex: | | |
|---|---|---|---|
| | FCCH$\cdots$H–Li $r(\text{H}\cdots\text{H}) = 1.963$ Å | FCCH$\cdots$H–Be–H $r(\text{H}\cdots\text{H}) = 2.192$ Å | FCCH$\cdots\sigma(\text{H}^2)$ $r(\text{H}\cdots\text{H}) = 2.595$ Å |
| ES | −7.39 | −1.68 | −0.32 |
| EX | 6.50 | 1.63 | 0.40 |
| PL | −2.6 | −0.37 | −0.05 |
| CT | −2.94 | −0.53 | −0.11 |
| Mix | 2.81 | 0.36 | 0.05 |

[a]ES, electrostatic; EX, exchange; PL, polarization; CT, charge transfer; Mix, mixing.

## 6.3. GROUP 3A: DIHYDROGEN BONDS X–H$\cdots$H–B, X–H$\cdots$H–Al, AND X–H$\cdots$H–Ga [X = FCC, HCC, LiCC, $CH_3OH$, $Pr^iOH$, $CF_3OH$, $CF_3CH_2OH$, $CFH_2CH_2OH$, $(CF_3)_2CHOH$, $(CF_3)_3COH$, CN, $CH_3$, INDOLE, IMIDAZOLE, PYRROLE, FKr, FAr]

The elements of group 3A—B, Al, and Ga—have a very important advantage over some other elements. On the one hand, hydrides of these elements are not very reactive compared with the elements of groups 1A and 2A, and on the other hand, they are not as complex from a computing point of view. In other words, they permit the use of theoretical approaches as well as experimental techniques for the investigation of dihydrogen bonding.

Among these elements, hydrides of which are capable of dihydrogen bonding, boron has been studied most intensively. The appearance of numerous experimental and theoretical works, focused, first, on borane amines, is explained by the following circumstances. First, borane amines, and compound $NH_3BH_3$ particularly, represent potential materials for storing and delivering large amounts of molecular hydrogen through dehydrogenation reactions. Therefore, studies of these compounds and their structural features are interesting practically. Second, atoms of boron do not have nonbonding valence electrons which could potentially participate in bonding to proton-donor molecules. In other words, the boron hydrides are the best models to obtain a clear understanding of the nature of dihydrogen bonding. Third, it is very easy to apply a variety of experimental techniques for the characterization of boron hydrides. In some sense these circumstances explain why the first experimental data on intermolecular proton–hydride interactions were obtained in the 1970s for $BH_3$ and $BH_2$ groups.

Calculation of the B, Al, and Ga hydrides and their dihydrogen-bonded complexes is not a problem for computers. Moreover, as we will see below, some of the approaches can be used for proper simulation of the effect of solvents on dihydrogen bonding. Nevertheless, in the absence of frequency analysis, the nature of a complex often remains unclear: Does it act as a transition state or as a minimum on the potential energy surface? One of the simplest examples is the dimer $(BH_3NH_3)_2$, investigated in numerous theoretical works. According to the data of Popelier, this dimer is formed due to the presence of three dihydrogen bonds. These bonds result in the structure shown in Figure 3.9, where the $H\cdots H$ bonding energy is determined as $-3.8$ kcal/mol and the nature of bonding is confirmed completely by the AIM criteria. However, as shown later by frequency analysis [18], in reality this structure represents a transition state. Two other structures, depicted in Figure 6.10, actually occupy the local minima on the potential energy surface.

It is interesting that the structure in the right in Figure 6.10 corresponds to the dimer, optimized earlier by Richardson and co-workers [19]. Nevertheless, the structure on the left, which has a total interaction energy of 12.8 kcal/mol (with a ZPE correction, the energy is equal to 11.2 kcal/mol), is more thermodynamically stable than Crabtree's dimer. At the same time, the difference in energy between the left- and right-side structures is very small, having been determined to be only 0.3 kcal/mol. These data show clearly the flatness of the potential energy surface, where one structure can easily be transformed to another. It is worth mentioning that the structural parameters and even the $H\cdots H$ distances, optimized for the ground-state structures and the transition states, differ, but not dramatically. This circumstance provides a mutual comparison.

Since many experimental and theoretical aspects of the $B-H\cdots H-X$ dihydrogen bonding have been discussed in earlier chapters, here we focus on a comparison of the parameters that characterize the B, Al, and Ga hydrides as proton acceptors. The study of interactions between the ions $[BH_4]^-$, $[AlH_4]^-$, and $[GaH_4]^-$ and acidic alcohols seems to be very interesting and representative in this context [20]. The dihydrogen-bonded complexes of these ions, which can be assigned to charge-assisted systems, have been found by computing at the B3LYP/6-311++G(d,p) and MP2/6-311++G(d,p) levels. One of the

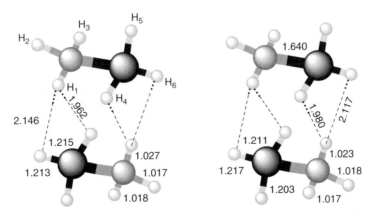

**Figure 6.10** Structures of dimers $(NH_3BH_3)_2$ located at the energy minima of the potential energy surface calculated at the DFT/B3LYP level. The dark spheres are the nitrogen atoms. (Reproduced with permission from ref. 18.)

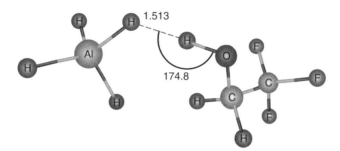

**Figure 6.11** Fully optimized geometry of the dihydrogen-bonded complex formed between $[AlH_4]^-$ and $CF_3CH_2OH$.

fully optimized geometries obtained for the $[AlH_4]^- \cdots HOCH_2CF_3$ complex is depicted in Figure 6.11, where, again, the H$\cdots$H–O moiety is practically linear. The H$\cdots$H distances in these complexes are remarkably shorter than the sum of the van der Waals radii of H (Table 6.9). The interaction energies obtained for the complexes are dominated by the electrostatic contributions given in Table 6.10. However, again, the polarization and charge transfer contributions are calculated to be significant compared to classical hydrogen bonds.

Table 6.10 shows the electron density parameters calculated for the complexes. As shown, they support completely the formulation of H$\cdots$H contacts as dihydrogen bonds. In fact, the bond critical points, located in the H$\cdots$H directions in complexes of the ions $[BH_4]^-$, $[AlH_4]^-$, and $[GaH_4]^-$ with methanol, are characterized by the small $\rho_C$ values of 0.025 to 0.026 au. In addition, the maximum $\rho_C$ values, varying between 0.053 and 0.062 au, have been calculated for those complexes with the more acidic alcohol, $CF_3OH$. Nontrivial values have been obtained for the Laplacian of the electron density, $\nabla^2 \rho_C$. They were negative in all the complexes noted above, indicating the presence of a slight covalency along the H$\cdots$H direction. Similar results were discussed earlier for other charge-assisted dihydrogen bonds.

According to the data in Table 6.9, the energies $\Delta E$ computed for dihydrogen-bonded complexes of the ions $[AlH_4]^-$ and $[GaH_4]^-$ are very similar. In fact, the difference is not larger than 0.4 kcal/mol for complexes of $[AlH_4]^-$ and $[GaH_4]^-$ with $CH_3OH$ or $CF_3OH$. At the same time the energies calculated for the ion $[BH_4]^-$ are notably higher, by $\approx 2$ kcal/mol. However, this tendency does not correlate with the H$\cdots$H distances, which decrease from B to Ga and then to Al. In addition, simulating the presence of THF reduces the energy difference. The nature of the effects observed is still unclear. Moreover, the interaction energies, obtained experimentally and collected in Table 6.11, also demonstrate another tendency. Here the $[GaH_4]^-$ dihydrogen complexes appear to be slightly stronger than the complexes of $[BH_4]^-$. On the other hand, the theory predicts accurately the experimentally observed influence of the proton-donor strength: a stronger donor corresponding to a stronger dihydrogen bond, in accordance with a shorter H$\cdots$H distance.

**TABLE 6.9. BSSE-Corrected Interaction Energies and H···H Distances for Dihydrogen-Bonded Complexes Formed by Hydrides of Elements in Group 3A**[a]

| Number | System | $-\Delta E$ (kcal/mol) | $r$(H···H) (Å) | Ref. |
|--------|--------|------------------------|----------------|------|
| 1 | $BH_4^-$···$HOCH_3$ | 12.5 (4.2)[b] | 1.654 | 20 |
| 2 | $AlH_4^-$···$HOCH_3$ | 10.5 (4.0)[b] | 1.622 | 20 |
| 3 | $GaH_4^-$···$HOCH_3$ | 10.3 (3.7)[b] | 1.628 | 20 |
| 4 | $BH_4^-$···$HOCF_3$ | 24.6 (8.9)[b] | 1.351 | 20 |
| 5 | $AlH_4^-$···$HOCF_3$ | 21.6 (8.5)[b] | 1.258 | 20 |
| 6 | $GaH_4^-$···$HOCF_3$ | 21.2 (7.7)[b] | 1.269 | 25 |
| 7 | $H_3BNH_3$···$H_3B-NH_3$ | 4.26[c] | 1.953 | 25 |
| 8 | $H_2BNH_2$···$H_2B-NH_2$ | 1.28[c] | 2.218 | 26 |
| 9 | $(CH_3)_3NBH_3$···$(CH_3)_3NBH_3$ | 5.26 | 2.13 | 26 |
|   |   |   | 2.31 |   |
| 10 | $(NH_3BH_3)_2$ | 12.8 | 1.952 | 18 |
| 11 | $(NH_3BH_3)_2$ | 15.1[c] | 1.990 | 27 |
| 12 | $(NH_3AlH_3)_2$ | 11.8[c] | 1.781 | 27 |
| 13 | $(NH_3GaH_3)_2$ | 10.7[c] | 1.898 | 27 |
| 14 | $FArH$···$HAlH_2$ | 7.08 | 1.085 | 15 |
| 15 | $FArH$···$HGaH_2$ | 5.64 | 1.070 | 15 |
| 16 | $FKrH$···$HAlH_2$ | 1.70 | 1.489 | 15 |
| 17 | $FKrH$···$HGaH_2$ | 1.5 | 1.472 | 15 |
| 18 | $BH_3$···$HCCF$ | — | 2.383 | 11 |
| 19 | $BH_3$···$HCCLi$ | — | 2.506 | 11 |
| 20 | $BH_3$···$HCCH$ | — | 1.802 | 11 |
| 21 | $[BH_4]^-$···$HCCLi$ | — | 2.086 | 11 |

[a]Obtained at the MP2/6-311++G* * level.
[b]Calculated by modeling the THF environment.
[c]Calculated at the MP2/aug-cc-pVTZ level.

Similar effects of proton-donor strength were found in a theoretical study of intermolecular interactions of $NH_3BH_3$ [21]. The structures of these $NH_3BH_3$ complexes are depicted in Figure 6.12, and the corresponding energies and H···H distances are given in Table 6.12. As shown, the very weak acid $CH_4$ yields an adduct with the very long H···H distances 2.467 and 2.929 Å. It is obvious that these distances are too long, and the interaction energy negligibly small, to formulate this structure as a dihydrogen–bonded complex. However, progressing to more pronounced proton donors such as $H_2O$, HCN, $CH_3OH$, and HF, the distances decrease and the interaction energies increase. Exceptions are structures S3 and S4, the former being a hydrogen-bonded complex and the high energy level of S4 being connected with multiple H···H contacts.

In contrast with the $CH_4$ molecule, the $NC–H^{\delta+}$ bonds in borane amine $(CH_3)_3NBH_3$ are slightly polarized. Even such slight polarization is sufficient for the formation of the dimer $(CH_3)_3NBH_3$···$(CH_3)_3NBH_3$, which can be formulated as a dihydrogen-bonded complex. Here, two bifurcated CH···H–B

**TABLE 6.10. Energy Contributions to the Total Energy of Three Dihydrogen-Bonded Complexes with Methanol and Electronic Density Parameters in the H$\cdots$H Directions**[a]

| | Complex | | |
|---|---|---|---|
| Parameter | $[BH_4]^- \cdot CH_3OH$ | $[AlH_4]^- \cdot CH_3OH$ | $[GaH_4]^- \cdot CH_3OH$ |
| $\rho_C$ (au) | 0.025 | 0.026 | 0.026 |
| $\nabla^2 \rho_C$ (au) | −0.014 | −0.013 | −0.012 |
| $E_{ES}$ (kcal/mol) | −12.5 | −11.3 | −11.7 |
| $E_{EX}$ (kcal/mol) | +5.7 | +5.8 | +7.5 |
| $E_{PL}$ (kcal/mol) | −3.5 | −2.4 | −3.9 |
| $E_{CT}$ (kcal/mol) | −2.0 | −2.0 | −3.7 |
| $E_{mix}$ (kcal/mol)[b] | +1.7 | +1.2 | +1.0 |
| $E_{tot}$ (kcal/mol) | −10.7 | −8.7 | −10.4 |

[a]Calculated using the 6-31++G** basis set.
[b]$E_{mix}$, the mixing term, represents the fact that $E_{ES}$, $E_{EX}$, $E_{PL}$, and $E_{CT}$ are not strictly independent of each other.

bonds with the H$\cdots$H distances of 2.13 and 2.31 Å give an energy gain of 5.25 kcal/mol (see Table 6.9).

Finally, the influence of the proton-accepting strength on dihydrogen bonding is shown clearly by the Al hydride complex data in Table 6.13. According to calculations, the energy gain increases from $AlF_2H$ to $AlH_3$, which are dihydrogen-bonded to HArF or HKrF. Accordly, the H$\cdots$H distances are reduced remarkably.

Among the dihydrogen-bonded systems presented in Table 6.9, two complexes, $H_3BNH_3\cdots H_3B-NH_3$ and $H_2BNH_2\cdots H_2B-NH_2$, are particularly interesting because they illustrate the effect of nitrogen and boron hybridization on dihydrogen bonding. As shown, these complexes exhibit H$\cdots$H distances of 1.953 and 2.218 Å, respectively, and their formation energy is calculated as 4.26 and 1.28 kcal/mol. Thus, the dihydrogen bond is significantly weaker in the system that has $sp^2$ nitrogen and boron hybridization.

Comparing the interaction energies obtained for the dihydrogen-bonded complexes of $[BH_4]^-$ and $[GaH_4]^-$ ions with the various proton donors in Tables 6.9 and 6.11 reveals a dramatic difference between the theoretical and experimental magnitudes. Since conventional theoretical investigations always imply the gas phase, this difference could be connected with the influence of the environment effective in the condensed phase. In accordance with this statement, the solvent effect of THF ($\varepsilon = 7.58$), simulated theoretically in the framework of the polarizable conductor calculation model, led to strongly reduced energies (see cases 1 to 6 in Table 6.9). Now the values calculated correspond more closely to experiments performed in $CH_2Cl_2$ solutions (see Table 6.11). According to the full optimizations of the complexes, these large differences in energies are associated with remarkable structural changes in the presence of THF. For example,

**TABLE 6.11. Experimental Interaction Energies Determined by IR Spectra for Dihydrogen-Bonded Complexes Formed by Hydrides of Group 3A Elements**[a]

| Number | System | $-\Delta E$ (kcal/mol) | Solvent | Ref. |
|---|---|---|---|---|
| 1 | $BH_4^-\cdots HOCH(CF_3)_2$ | 6.5 | $CH_2Cl_2$ | 28 |
| 2 | $BH_4^-\cdots HOCH_2CF_3$ | 5.2 | $CH_2Cl_2$ | 28 |
| 3 | $BH_4^-\cdots HOCH_3$ | 4.1 | $CH_2Cl_2$ | 28 |
| 4 | $BH_4^-\cdots HOPr$ | 3.8; 3.2[b] | $CH_2Cl_2$ | 28 |
| 5 | $BH_4^-\cdots$indole | 2.5 | $CH_2Cl_2$ | 28 |
| 6 | $[B_{10}H_{10}]^{2-}\cdots HOC(CF_3)_3$ | 4.2 | $CH_2Cl_2$ | 29 |
| 7 | $[B_{10}H_{10}]^{2-}\cdots HOCH(CF_3)_2$ | 4.1 | $CH_2Cl_2$ | 29 |
| 8 | $[B_{10}H_{10}]^{2-}\cdots HOCH_2CF_3$ | 3.3; 3.2[b] | $CH_2Cl_2$ | 29 |
| 9 | $[B_{10}H_{10}]^{2-}\cdots HOCH_3$ | 2.6 | $CH_2Cl_2$ | 29 |
| 10 | $[B_{10}H_{10}]^{2-}\cdots HOPr$ | 2.3 | $CH_2Cl_2$ | 29 |
| 11 | $[B_{12}H_{12}]^{2-}\cdots HOCH(CF_3)_2$ | 3.2 | $CH_2Cl_2$ | 29 |
| 12 | $[B_{12}H_{12}]^{2-}\cdots HOCH_2CF_3$ | 2.4; 2.2[b] | $CH_2Cl_2$ | 29 |
| 13 | $[B_{12}H_{12}]^{2-}\cdots HOCH_2CH_3$ | 1.9 | $CH_2Cl_2$ | 29 |
| 14 | $[B_{12}H_{12}]^{2-}\cdots HOPr$ | 1.8 | $CH_2Cl_2$ | 29 |
| 15 | $BH_4^-\cdots HCN$ | 17.02 | MP2/6- | 5 |
| 16 | $BH_4^-\cdots H_4C$ | 1.2 | 311++G** | 5 |
| 17 | $GaH_4^-\cdots HOCH_2CF_3$ | 5.4 | $CH_2Cl_2$ | 24 |
| 18 | $GaH_4^-\cdots HOCH2CH_2F$ | 4.4 | $CH_2Cl_2$ | 24 |
| 19 | $GaH_4^-\cdots$indole | 4.4 | $CH_2Cl_2$ | 24 |
| 20 | $GaH_4^-\cdots HOCH_3$ | 4.3 | $CH_2Cl_2$ | 24 |
| 21 | $GaH_4^-\cdots HOPr$ | 4.0 | $CH_2Cl_2$ | 24 |
| 22 | $Et_3NBH_3\cdots HOCH(CF_3)_2$ | 3.7 | $CH_2Cl_2$ | 28 |
| 23 | $Et_3NBH_3\cdots HOCH(CF_3)_2$ | 3.5; 3.4[b] | Hexane | 28 |
| 24 | $Et_3NBH_3\cdots HOCH_2CF_3$ | 2.7 | Hexane | 28 |
| 25 | $Et_3NBH_3\cdots HOCH_3$ | 1.9 | Hexane | 28 |
| 26 | $Et_3NBH_3\cdots HOPr$ | 1.7 | Hexane | 28 |
| 27 | $Et_3NBH_3\cdots$indole | 1.3 | Hexane | 28 |
| 28 | $Et_3OPBH_3\cdots HOC(CF_3)_3$ | 3.6 | Hexane | 28 |
| 29 | $Et_3OPBH_3\cdots HOCH(CF_3)_2$ | 2.5 | Hexane | 28 |
| 30 | $Et_3OPBH_3\cdots HOCH_2CF_3$ | 1.9 | Hexane | 28 |
| 31 | $Et_3OPBH_3\cdots HOPr$ | 1.1 | Hexane | 28 |
| 32 | $[BH_3CN]^-\cdots HOCH(CF_3)_2$ | 3.8 | $CH_2Cl_2$ | 30 |
| 33 | $[BH_3CN]^-\cdots HOCH_2CF_3$ | 3.6 | $CH_2Cl_2$ | 30 |
| 34 | $[BH_3CN]^-\cdots HOCH_2CH_2F$ | 3.2 | $CH_2Cl_2$ | 30 |
| 35 | $[BH_3CN]^-\cdots$imidazole | 2.8 | $CH_2Cl_2$ | 30 |
| 36 | $[BH_3CN]^-\cdots HOCH_3$ | 2.3 | $CH_2Cl_2$ | 30 |
| 37 | $[BH_3CN]^-\cdots$pyrrole | 2.2 | $CH_2Cl_2$ | 30 |

[a]Data obtained on the basis of the $\nu(OH)$ or $\nu(NH)$ frequency region.
[b]Determined directly as the $\Delta H^\circ$ value by variable-temperature IR experiments where $\Delta S^\circ$ was between $-3.8$ and $-6.0$ eu.

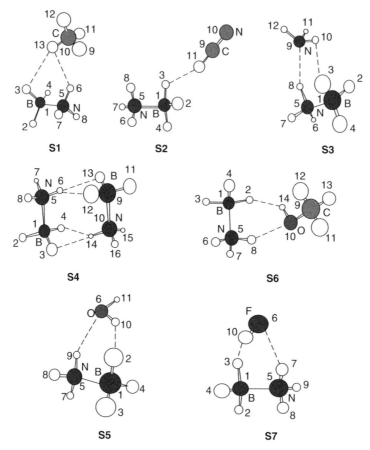

**Figure 6.12** Geometry of the complexes formed by $NH_3BH_3$ and $CH_4$ (**S1**), HCN, (**S2**), $NH_3$ (**S3**), $NH_3BH_3$ (**S4**), $H_2O$ (**S5**), $CH_3OH$ (**S6**), and HF (**S7**) as proton donors. (Reproduced with permission from ref. 21.)

the H···H distance in the dihydrogen-bonded systems $GaH_4^-$···$H–OCF_3$ and $AlH_4^-$···$H–OCF_3$ elongates in the presence of THF by 0.097 and 0.088 Å, respectively, whereas the O–H and metal–H distances become shorter. Strong effects of cyclohexane, water, and DMSO on the geometry of the $NH_3BH_3$···HF complex have been also modeled by Meng and co-workers [21].

Compounds $BH_3NH_3$, $AlH_3NH_3$, and $GaH_3NH_3$, containing proton-donor bonds N–H and proton acceptor bonds B–H, spontaneously give the dimers $(BH_3NH_3)_2$, $(AlH_3NH_3)_2$, and $(GaH_3NH_3)_2$, built via dihydrogen bonds. These dihydrogen-bonded dimers play an important role in understanding the solid-state structure of B, Al, and Ga hydrides, particularly their solid-state dynamics. For example, the influence of dihydrogen bonds on the $BH_3$ and $NH_3$ rotor motions in solid $BH_3NH_3$ has been found by $^{15}N$, $^1H$, $^2H$, and $^{11}B$ solid-state NMR

**TABLE 6.12. H· · ·H Distances and Interaction Energies for Dihydrogen-Bonded Systems S1 to S7 in Figure 6.12**[a]

| Structure | H· · ·H (Å) | $-\Delta E$ (kcal/mol) | $-\Delta E_{BSSE}$[b] (kcal/mol) |
|---|---|---|---|
| S1 | 2.467 | 0.43 | 0.43 |
|  | 2.929 |  |  |
| S2 | 2.211 | 4.45 | 4.45 |
| S3 | 2.662 | 8.30 | 8.35 |
| S4 | 1.898 | 12.63 | 12.63 |
|  | 1.965 |  |  |
|  | 2.238 |  |  |
|  | 2.126 |  |  |
| S5 | 1.929 | 8.11 | 8.04 |
| S6 | 1.956 | 8.23 | 8.11 |
| S7 | 1.503 | 9.62 | 9.62 |

[a]Calculated at the B3LYP/6-311++G(d,p) level.
[bb]Values after BSSE corrections.

**TABLE 6.13. Bonding Energies and H· · ·H Distances for Dihydrogen-Bonded Complexes of Al Hydrides**[a]

| Number | System | $-\delta E$ (kcal/mol) | r(H·H) (Å) |
|---|---|---|---|
| 1 | $H_2AlH· · ·HArF$ | 8.74 | 1.085 |
| 2 | $ClHAlH· · ·HArF$ | 5.52 | 1.269 |
| 3 | $FHAlH· · ·HArF$ | 4.92 | 1.320 |
| 4 | $Cl_2AlH· · ·HArF$ | 3.17 | 1.414 |
| 5 | $F_2AlH· · ·HArF$ | 1.68 | 1.556 |
| 6 | $H_2AlH· · ·HKrF$ | 3.12 | 1.489 |
| 7 | $ClHAlH· · ·HKrF$ | 1.75 | 1.616 |
| 8 | $FHAlH· · ·HKrF$ | 1.65 | 1.651 |
| 9 | $Cl_2AlH· · ·HKrF$ | 0.95 | 1.735 |
| 10 | $F_2AlH· · ·HKrF$ | 0.17 | 1.888 |

*Source:* Ref. 22.
[a]Calculated at the MP2/6-311++G** level.

spectra [23]. On the other hand, these dimers can have potential applications in crystal engineering, where the effects connected with the nature of an element are particularly important. Since the dimers $(BH_3NH_3)_2$, $(AlH_3NH_3)_2$, and $(GaH_3NH_3)_2$ are structurally similar, the nature of the element can be traced by comparing the energies of dimerization. The data in Table 6.9 show, first, that this energy is significant, and second, that it changes in the order B > Al ≥ Ga. It is interesting that the H· · ·H distance shows another order, B > Ga > Al, as we observed earlier.

Most of the bonding energies in Table 6.11 have been determined by IR spectra on the basis of the frequency shifts in the the ν(OH) or ν(NH) regions

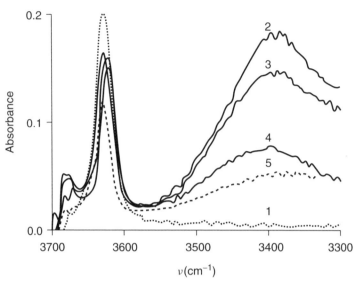

**Figure 6.13** IR spectra recorded in the $\nu(OH)$ region of $CH_3OH$ in a $CH_2Cl_2$ solution at 260 K (1) and in the presence of [BuN][GaH$_4$] at 200 K (2), 220 K (3), 260 K (4), and 290 K (5). (Reproduced with permission from ref. 24.)

via the relationship in eq. (4.4). However, some of them were measured directly in variable-temperature IR experiments. One of the experiments is illustrated in Figure 6.13, where the $\nu(OH)$ band of $CH_3OH$ undergoes a red shift in the presence of the ion [GaH$_4$]$^-$ and the intensity of the new wideband decreases with increased temperature (see the spectra from 2 to 5) [24]. It is obvious that these spectral changes are caused by a shift of equilibrium,

$$[GaH_4]^- + CH_3OH \rightleftharpoons [GaH_4]^- \cdots HO-CH_3 \qquad (6.5)$$

toward the dihydrogen-bonded complex, whose formation constants can be determined easily at each temperature. Then their temperature dependencies lead to $\Delta H°$ values which are reasonably smaller than $\Delta E$.

To conclude this section it is should be emphasized that the effects of the proton-donor and proton-acceptor strengths on bonding energies, predicted theoretically, are clear from the experimental data. In fact, it follows from Table 6.11 that the energy gain is maximal for the pairs [BH$_4$]$^-$/HOCH(CF$_3$)$_2$ and [GaH$_4$]$^-$/HOCH(CF$_3$)$_2$ (6.5 and 5.4 kcal/mol, respectively) and minimal in the case of the Et$_3$OPBH$_3$···HOPr system (1.1 kcal/mol). The latter is formed by boron hydride with a small hydridicity of hydrogen atoms, due to the presence of the electron-attractive Et$_3$OP group.

## 6.4. HYDRIDES OF THE ELEMENTS IN GROUP 4A ACTING AS PROTON ACCEPTORS AND PROTON DONORS: VERY WEAK DIHYDROGEN BONDS

A priori, the bonds C–H, Si–H, and Ge–H are almost covalent and their polarization is minimal (here we do not mean the specific class of strong C–H acids, capable of hydrogen bonding, etc.). For this reason, they can play the role of a very weak proton donor or proton acceptor, forming dihydrogen bonds with small bonding energies. This is particularly valid for intermolecular interactions C–H$\cdots$H–C, where the electrostatic component is completely absent. In accord with this statement, an analysis of the Cambridge Structural Database for organic molecules containing methyl or ethyl groups has demonstrated that such C–H$\cdots$H–C contacts are characterized by the ideally isotropic angular distribution typical of van der Waals interactions [31]. However, as we will see in this chapter, even C–H$\cdots$H–C contacts can sometimes have a bonding character.

### 6.4.1. Bonds C–H, Si–H, and Ge–H in Dihydrogen Bonding

Historically, the first complexes, formed by the strong acids HCl, HF, HBr, and HCN, on the one hand, and cyclopropane or $CH_4$ molecules, on the other, have been detected in the gas phase with the help of the matrix isolation microwave spectroscopy. The complexes were reviewed in 1983 by Barnes [32] and later by Legon and co-workers [33]. It is interesting because it was done long before the appearance of the term *dihydrogen bond*. This matrix technique provides a pseudo-gas-phase environment for weak molecular complexes, which results in much sharper bands with respect to those observed in the gas phase or solutions. In addition, under matrix conditions, complexes that usually react in the gas phase can be stabilized and studied. Figure 6.14 illustrates this advantage showing IR spectra where the bands observed are actually narrow. These low-temperature IR spectra have been recorded for HCl placed in an argon matrix in the absence and presence of methane, added to HCl in the ratio $1:1$. As shown, $CH_4$ causes significant spectral perturbations interpreted by the formation of two complexes, marked C (the $1:1$ complex) and C' (the $1:2$ complex, formed by HCl as a dimer). The complexes have been classified as hydrogen bonded on the basis of the definition that the hydrogen atom of the $CH_4$ molecule is located between the Cl and C atoms. Similar results have been reported for the cyclopropane·HCl complex, interpreted, however, in terms of an edge-bonded structure (Figure 6.15). Computational investigations of $CH_4$ complexes [34] have confirmed the conclusion that the $CH_4$ molecule behaves as a proton acceptor.

Later, Atkins and co-workers [35] studied the rotational spectra of $CH_4$·HX systems using a pulse-nozzle-FT-microwave spectrometer operating in the region 4 to 18 GHz. The rotational transition frequencies observed have allowed us to deduce the equilibrium geometries of $CH_4$·HX complexes. Figure 6.16 shows the geometry of two of them, where X = HCl and HCN. The complexes have $C_3$ symmetry, and a hydrogen bond is directed to the center of a face of the $CH_4$ molecule.

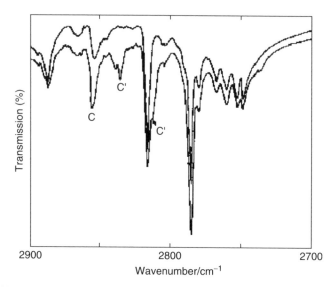

**Figure 6.14** IR spectrum of HCl recorded in an argon matrix at an HCl/Ar ratio of 1 : 200 (top) and IR spectrum of a $CH_4$/HCl/Ar mixture with a ratio 1 : 1 : 200 (bottom). (Reproduced with permission from ref. 32.)

Another equilibrium geometry has been suggested for the $CH_4 \cdot HF$ system (Figure 6.17). Here, two hydrogen-bonding interactions take place rather than just one, resulting in remarkable deviation from $C_{3v}$ symmetry. On the other hand, this deviation could be interpreted in terms of dihydrogen bonding, which is responsible for the proximity of the C–H and H–F hydrogen atoms.

The first high-level theoretical evidence for the existence of dihydrogen bonds between $CH_4$ and $[NH_4]^+$ came in 1996 [5]. In the sense of bonding geometry, this complex is not linear (Structure 6.5), has $C_{3v}$ symmetry, and exhibits three relatively short H$\cdots$H contacts with distances of 2.237, 2.357, and 2.226 Å calculated at the RHF/6-31G**, RHF/6-311++G**, and MP2/6-31G** levels, respectively. According to the frequency analysis at the RHF/6-31G** level, the complex occupies a minimum on the potential energy surface of this system. The BSSE-corrected interaction energies have been calculated as −2.46,

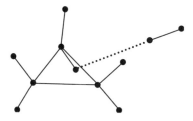

**Figure 6.15**

−2.52, and −3.51 kcal/mol at the same theoretical levels. It should be noted that −3.51 kcal/mol calculated at the MP2/6-31G** level agrees well with the experimental $\Delta H°$ value of −3.59 kcal/mol determined for formation of the $[NH_4 \cdot CH_4]^+$ complex in the gas phase [36]. Topological analysis of the electron density performed in the framework of AIM theory shows the bond critical points on the H· · ·H directions with $\rho_C$ values of 0.013 au. It is interesting that the electron density in this complex is larger than that obtained for the $BH_4^- \cdot \cdot \cdot CH_4$ dihydrogen-bonded system ($\rho_C = 0.007$ au), the $CH_4$ molecule of which acts as a proton donor. In accordance with the electronic density, the H· · ·H distances in the $BH_4^- \cdot \cdot \cdot H_4C$ complex were remarkably longer than 2.4 Å (2.797, 2.929,

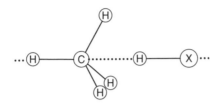

**Figure 6.16**

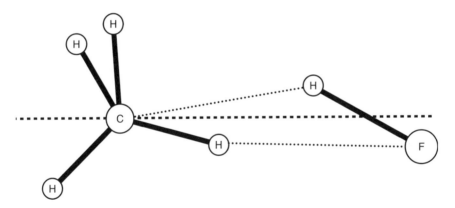

**Figure 6.17**

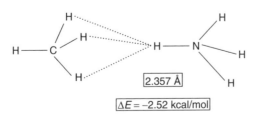

**Structure 6.5** Geometry and interaction energy calculated for the dihydrogen complex $CH_4 \cdot [NH_4]^+$ at the RHF/6-311++G** level.

**TABLE 6.14. Structural and Energy Parameters Calculated at Various Theoretical Levels for $XH_4 \cdots HF$ Complexes of Carbon, Silicon, Germanium, and Tin**

| $XH_4$ | $-\Delta E$ (kcal/mol) | $r(H \cdots H)$ (Å) | H-H | Method |
|---|---|---|---|---|
| $CH_4$ | −0.27 | 1.982 | 0.949 | DFT/B3LYP |
| | +0.39 | 2.613 | 0.917 | MP2/6-311++G** |
| | +0.12 | 2.471 | 0.921 | MP2/6-31G** |
| $SiH_4$ | −1.69 | 1.738 | 0.953 | DFT/B3LYP |
| | −0.85 | 1.947 | 0.919 | MP2/6-311++G** |
| | −0.65 | 2.045 | 0.922 | MP2/6-31G** |
| $GeH_4$ | −1.95 | 1.716 | 0.954 | DFT/B3LYP |
| $SnH_4$ | −2.78 | 1.643 | 0.956 | DFT/B3LYP |

*Source:* Data from refs. 6 and 10.

and 2.583 Å at the RHF/6-31G**, RHF/6-311++G**, and MP2/6-31G** levels, respectively), corresponding to small interaction energies [5]. In this connection, the nature of this complex is closer to that of van der Waals interactions.

In contrast to the $CH_4 \cdot [NH_4]^+$ complex, where the $CH_4$ molecule acts as a three-center proton acceptor, the geometry of the dihydrogen complex between $CH_4$ and HF, optimized by Grabowski [6] at the MP2/6-31G*, MP2/6-311 ++G**, and MP4/6-311++G** levels, is linear. In addition, it shows a single $H \cdots H$ contact calculated as 2.471 and 2.613 Å at the MP2/6-31G* and MP2/6-311++G** levels, respectively. Similar results have been reported for the complexes HF · $SiH_4$, HF · $GeH_4$, and HF · $SnH_4$, the structural and energy parameters of which are shown in Table 6.14. As shown, these parameters depend heavily on the computing level. For example, complex $CH_4 \cdots HF$ is stable only at the DFT/B3LYP level. Nevertheless, the data indicate two important tendencies: The complexes are formed due to the relatively short $H \cdots H$ contacts, and the formation of the complexes is accompanied by a slight elongation of the H−F bond since the $H \cdots F$ bond length in the isolated H−F molecule is calculated as 0.917 Å. For comparison, the table shows the parameters for the $SnH_4 \cdots HF$ dihydrogen-bonded complex where the bonding energy is highest, the $H \cdots H$ distance is shortest, and the effect of the H−F bond elongation is maximal.

AIM topological analysis of the electron density performed for two complexes and for isolated components is shown in Table 6.15. The bond critical points found in the $H \cdots H$ directions are characterized by the small electronic density with $\rho_C = 0.002$ and 0.009 au in the $CH_4 \cdots HF$ and $SiH_4 \cdots HF$ systems, respectively. The Laplacian, $\nabla^2 \rho_C$, is also small but takes positive values in accordance with the AIM criteria for dihydrogen bonding.

Comparing the data in Tables 6.14 and 6.15 shows that dihydrogen bonding to the same proton donor depends strongly on the nature of the group 3A elements: The interaction energy increases in the order

$$C < Si < Ge < Sn \qquad (6.6)$$

**TABLE 6.15. Electron Density Analysis for $XH_4\cdots HF$ Complexes and Isolated Monomers $XH_4$ and $HF^a$**

| System | $\rho_C$ (au) | $\nabla^2\rho_C$ (au) |
|---|---|---|
| $H_3C-H$ | 0.272 | $-0.912$ |
| $H_3Si-H$ | 0.119 | 0.287 |
| $H-F$ | 0.370 | $-0.284$ |
| $H_3C-H\cdots H-F$ | 0.270 (C–H) | $-0.884$ (C–H) |
| | 0.002 (H$\cdots$H) | 0.005 (H$\cdots$H) |
| | 0.370 (H–F) | $-2.830$ (H–F) |
| $H_3Si-H\cdots H-F$ | 0.115 (Si–H) | 0.285 (Si–H) |
| | 0.009 (H$\cdots$H) | 0.028 (H$\cdots$H) |
| | 0.367 (H–F) | $-2.810$ (H–F) |

$^a$At the MP2/6-311++G** level within AIM theory.

In the case of carbon and silicon, this effect correlates with an increase in the $\rho_C$ value from 0.002 au to 0.009 au in the H$\cdots$H direction.

As noted above, the complex $CH_4\cdots HF$ has been found to be stable only at the DFT/B3LYP level, while MP2 calculations have shown too-long H$\cdots$H distances at small positive $\Delta E$ values (Table 6.14). In other words, this complex cannot be formulated as dihydrogen bonded in the framework of MP2 theory. Recently, Govender and Ford [37] have reinvestigated the systems $CH_4 \cdot HF$, $SiH_4 \cdot HF$, $CH_4 \cdot HCl$, and $SiH_4 \cdot HCl$ that have been calculated at the MP2/6-311++G(d,p) level. On the basis of the vibrational spectra, calculated for the complexes of methane, they have been reclassified as hydrogen-bonded adducts, containing bonds $XH\cdots C$. However, the systems formed by $SiH_4$ have again shown dihydrogen bonds.

The detailed theoretical study of dihydrogen bonding in the complexes between $SiH_4$ and $[NH_4]^+$ has been performed by Zhu and co-workers [38]. The authors have taken the B3LYP and MP2 approaches using different basis sets from 6-31G* to 6-311++G**. Figure 6.18 depicts the set of structures with different proton–hydride interactions that have been taken initially for further optimizations. It should be noted that structure (f) is very similar to Alkorta's structure in Structure 6.5 optimized for the complex $CH_4 \cdot [NH_4]^+$. However, in contrast to Alkorta's data, all of these initial structures relaxed to the same complex, regardless of the method and basis set used. This complex shows a single short H$\cdots$H contact at a linear geometry. The H$\cdots$H distance has been calculated as 1.717, 1.676, and 1.613 Å at the 6-31G*, 6-31G**, and 6-311++G(2df,2pd) levels, respectively. The bonding energy has been estimated as $-5.595$ (B3LYP/6-311++G(2df,2pd)) and $-4.465$ (MP2/6-311++G(d,2p)) kcal/mol after the BSSE corrections.

Finally, Pakiari and Mohajeri [39] analyzed the structures in Figure 6.18 as models of multidihydrogen bonds. According to the calculations, performed using the RFH, B3LYP, and MP2 methods, linear coordination is characterized by the shortest H$\cdots$H distance of 1.7 Å, in good agreement with previous data

[38]. Strongly elongated H···H bonds have been found for the other structures, changing from 2.134 and 2.990 Å, as a function of the structure and the method of calculation. Bonding energies, also dependent on the method of calculations, were between $-1.3$ and $-3.99$ kcal/mol, characterizing the dihydrogen bonds as weak.

BAs we have shown above, interpretation of weak interactions is difficult even in the case of the complex $CH_4 \cdot HX$, where HX is a pronounced proton donor. It is obvious that interactions $C-H \cdots H-C$ are more questionable in this sense. Nevertheless, a number of intermolecular $C-H \cdots H-C$ contacts, which are smaller than 3 Å, have been observed: for example, in the crystal structures of the organoammonium tetraphenylborate system [40]. Primarily, these H···H contacts are observed between hydrogen atoms in the phenyl rings located in neighboring

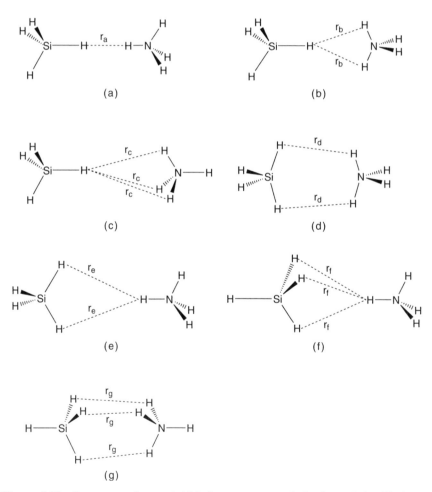

**Figure 6.18** Structures, taken as initial, for geometry optimizations of the dihydrogen complex formed by $SiH_4$ and $[NH_4]^+$. (Reproduced with permission from ref. 38.)

**TABLE 6.16. Structural Parameters and Characteristics of Bond Critical Points Found for C–H···H′–C′ Contacts in Tetraphenylborate, {[(H₂N)₂C–]₂N}[BPh₄]**

| Type of Contact | $r$(H···H) (Å) | Angle C–H···H′ (deg) | Angle H···H′–C′ (deg) | $\rho_C$ (e/Å³) | $\nabla^2\rho_C$ (e/Å⁵) |
|---|---|---|---|---|---|
| o-m | 2.179 | 165.8 | 103.1 | 0.078 | 0.703 |
| o-m | 2.205 | 175.5 | 117.3 | 0.056 | 0.583 |
| o-o | 2.261 | 115.1 | 115.1 | 0.057 | 0.782 |

BPh₄ anions. Robertson and co-workers have carefully analyzed these interactions using AIM theory [11], where the H···H contacts have been characterized by well-defined bond paths and bond critical points. Some of the geometric and electron density parameters obtained for tetraphenylborate {[(H₂N)₂C–]₂N}[BPh₄] are shown in Table 6.16. A careful analysis of the data obtained in terms of the AIM criteria for linear X–H···H–Y bonds has shown that the C–H···H–C contacts investigated can indeed be formulated as dihydrogen bonds. These bonds are weaker than the classical and nonclassical hydrogen bonds N–H···N, N–H···Ph, and C–H···Ph. Nevertheless, as we show below, interactions CH···HC can contribute to stabilization of crystals despite their energy weakness.

## 6.5. H–H BONDING

Matta and co-workers [41] have recently investigated a number of organic compounds, such as phenanthrene, tetra-*tert*-butyltetrahedrane, tetra-*tert*-butyl-cyclobutadiene, and tetra-*tert*-butylindacene that could potentially show interactions C–H···H–C. The B3LYP/6-31G* calculations have resulted in molecular graphs, one of which is shown schematically in Figure 6.19. This graph, built for the planar biphenyl molecule, exhibits bond critical points in the H···H directions. Topological analysis of the electron density has revealed the intramolecular H–H bond paths, and the bond critical points have been characterized by small $\rho_C$ values (from 0.0028 to 0.0168 au) and also small positive Laplcians, $\nabla^2\rho_C$ (from 0.0093 to 0.0535 au). It has been also found that the $\rho_C$ value deceases practically linearly with increasing distance between hydrogen atoms. Thus, according to the analysis, these interactions are closed shell, where the electron density values are typical of those found for weak hydrogen bonds and somewhat larger than those calculated for van der Waals interactions. In terms of AIM theory, these interactions can formally be assigned to dihydrogen bonds. However, Bader emphasizes that in contrast to dihydrogen bonds, as a part of hydrogen bonds, the interaction C–H···H–C occurs between two hydrogen atoms that are *identically or similarly charged* (see Section 2.1). In other words, *no electrostatic component dominates in this interaction*, which should be called, more accurately, *H–H bonding* or H–H interaction.

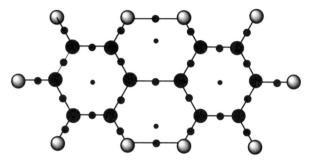

**Figure 6.19** Molecular graph (schematically) calculated for the planar biphenyl molecule at the B3LYP/6-31G* level.

On the basis of the data obtained, the three most important distinctions between *H–H bonding* and *dihydrogen bonding* are:

1. The lengths of the C–H bonds under H–H bonding *decrease* by 0.002 to 0.004 Å, whereas the opposite effect is typical of hydrogen or dihydrogen bonding (the well-known red shifts in vibrational frequencies).
2. Hydrogen bonding A–H···B causes the transfer of electronic charge from H to A and B by 0.01 to 0.2 e. Thus, the already large *positive* charge on the H atom *increases* additionally. Hydrogen atoms in the polybenzeneoids bear slight *negative* charges that *increase* in magnitude at the H–H bond.
3. Under hydrogen bonding, the energy of the acidic hydrogen increases and the destabilizing energy lies between 20 and 40 kcal/mol. H–H bonding results in stabilizing energy for the pair of atoms by up to 10 kcal/mol.

Matta and co-workers [41] conclude that stability of the molecules investigated is explained partially in terms of the energy that is necessary to disrupt the encasing network of these H–H bond paths. These interactions must be ubiquitous, and their stabilization energies contribute to the sublimation energies of hydrocarbon molecular crystals.

## 6.6. XENON DIHYDROGEN-BONDED COMPLEXES

Xe and Kr derivatives of the type H–Xe–Y and H–Kr–Y, where Y = H, CN, SH, or F, have attracted considerable attention from chemists in recent years. Some of these compounds show a partial negative charge on the hydrogen atoms [42]. For example, the Mulliken charge in $XeH_2$ is calculated at the MP2/6-31G** level as −0.34 e, which is close to that in ions $BH_4^-$ (−0.27 e). Thus, the $XeH_2$ molecule is a good candidate for dihydrogen bonding.

Xenon dihydride is frequently observed in Xe matrices, which contain hydrogen atoms. In turn, in the context of applications of this matrix isolation technique, water is an impurity that is difficult to eliminate. For these reasons, dihydrogen

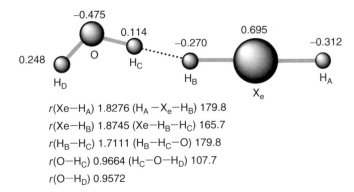

$r(Xe-H_A)$ 1.8276 ($H_A-Xe-H_B$) 179.8
$r(Xe-H_B)$ 1.8745 ($Xe-H_B-H_C$) 165.7
$r(H_B-H_C)$ 1.7111 ($H_B-H_C-O$) 179.8
$r(O-H_C)$ 0.9664 ($H_C-O-H_D$) 107.7
$r(O-H_D)$ 0.9572

**Figure 6.20** Structure of the dihydrogen-bonded complex formed by $XeH_2$ and, $H_2O$ optimized at the MP2/6-311++G(2 d,2p) level. The bond distances and bond angles are given in angstroms and degrees, respectively.

bonding between $XeH_2$ and $H_2O$ is particularly interesting. Such dihydrogen complexes have actually been localized on the potential energy surface of the systems $XeH_2 \cdot H_2O$ and $XeH_2 \cdot 2H_2O$ calculated at various theoretical levels [43–45].

According to the calculations, the linear centrosymmetric $XeH_2$ molecule interacts with one water molecule to give a dihydrogen-bonded complex with the planar geometry shown in Figure 6.20. It is remarkable that only one energy minimum has been located on the potential energy surface of the system. As can be seen, the complexation leads to a very short $H \cdots H$ distance and an elongated $Xe-H$ bond with respect to the free $Xe-H$ bond. The Mulliken charges decrease on both Xe hydrogens. Finally, depending somewhat on the method, the bonding energy shows that the complex is rather weak (Table 6.17).

In contrast to the $XeH_2 \cdots H_2O$ complex, the $XeH_2 \cdot 2H_2O$ system has shown three stationary points, corresponding to complexes **I**, **II**, and **III** in Figure 6.21.

**TABLE 6.17. Energy of a Dihydrogen Bond Between $XeH_2$ and a Water Molecule (Corrected with BSSE) Calculated at Various Theoretical Levels**

| Method | $-\Delta E$ (kcal/mol) |
|---|---|
| B3LYP/6-311++G(2 d, 2p) | 2.44 |
| B3LYP/6-311++G(3 d, 3p) | 2.31 |
| B3LYP/6-311++G(3df, 3p) | 2.29 |
| B3LYP/6-311++G(3df, 3dp) | 2.33 |
| B3LYP/aug-cc-pVTZ | 2.37 |
| MP2/6-311++G(2 d, 2p) | 2.53 |
| MP3/6-311++G(2 d, 2p)// MP2/6-311++G(2 d, 2p) | 2.13 |
| CCSD/6-311++G(2 d, 2p)// MP2/6-311++G(2 d, 2p) | 2.15 |

The first complex is nonplanar and exhibits two dihydrogen bonds, with the H$\cdots$H distances elongated to 1.942 Å (compare with 1.7111 Å found for the 1 : 1 complex). It is worth noting that this nonplanar structure occupies an energy minimum on the potential energy surface of the XeH$_2$ $\cdot$ 2H$_2$O system, whereas a completely planar structure represents a transition state. Dihydrogen-bonded complex **II** with a shorter H$\cdots$H distance of 1.756 Å is similar to the 1 : 1 complex. However, complex **II** is additionally stabilized by the classical hydrogen bond between H(3) and O(2). Structure **III**, a variation of **II**, is also stabilized by the additional classical hydrogen bond between O(1) and H(5). However, its H$\cdots$H distance is shorter (1.659 Å). For obvious reasons the total bonding energy increases from a 1 : 1 dihydrogen-bonded complex to the double dihydrogen-bonded system **I**, then to the dihydrogen/hydrogen-bonded complexes **II** and **III** (compare the data in Tables 6.17 and 6.18).

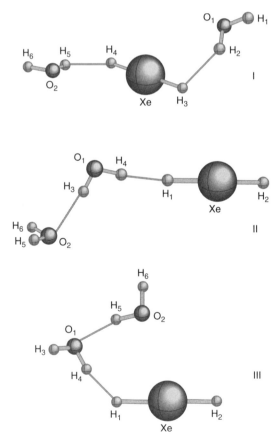

**Figure 6.21** Equilibrium structures of dihydrogen-bonded complexes formed by XeH$_2$ and 2H$_2$O optimized at MP2/6-311++G(2 d,2p). (Reproduced with permission from ref. 44.)

**TABLE 6.18. BSSE-Corrected Energy Obtained for Dihydrogen Bonding HXeH · (H₂O)₂ (see Figure 6.21)**

| Method | $-\Delta E$ (kcal/mol) for Complex: | | |
| | I | II | III |
| --- | --- | --- | --- |
| MP2 | 5.49 | 6.73 | 8.67 |
| MP3 | 4.76 | 6.30 | 7.86 |
| CCSD | 4.20 | 5.80 | 7.04 |

## 6.7. CONCLUDING REMARKS

1. Dihydrogen bonds Li–H· · ·H–X and Na–H· · ·H–X are generally linear and change from weak (4.3 kcal/mol) to very strong (27 kcal/mol) as a function of the strength of proton donors. In accord, the H· · ·H distance increases from 1.31 Å to 2.13 Å. On going from Li to Na, the interaction energy increases and the H· · ·H distance decreases. In contrast to numerous dihydrogen-bonded systems with significant predominance of the electrostatic interaction, for Li and Na there are some examples where the charge transfer contribution surpasses the electrostatic energy.

2. Simple dihydrogen bonds Be–H· · ·H–X and Mg–H· · ·H–X are also linear. Their energy decreases from 20.6 kcal/mol to 0.91 kcal/mol when the H· · ·H distance increases from 1.11 Å to 2.29 Å. Similar to the Li and Na dimer complexes, the geometry of the better associated Be and Mg systems is not linear and depends on the nature of proton donors. Dihydrogen-bonded complexes of Be and Mg are remarkably weaker than their Na and Li analogs.

3. Energy of Al–H· · ·H–X, Ga–H· · ·H–X, and B–H· · ·H–X dihydrogen bonds, computed for different proton donors, changes from 22 kcal/mol to 1.3 kcal/mol when the distance is lengthened from 1.26 Å to 2.22 Å. The energy values for the ions [AlH₄]⁻ and [GaH₄]⁻ are very similar, whereas they are remarkably higher for ions [BH₄]⁻. On going from the gas phase (theoretical data) to the condensed state (the IR data in solutions) the dihydrogen bonding energy decreases. It has been found that the bonding energy in the dimers (BH₃NH₃)₂, (AlH₃NH₃)₂, and (GaH₃NH₃)₂ changes in the order B > Al ≥ Ga. The H· · ·H distance shows another order: B > Ga > Al. The dihydrogen-bonded complexes of Al, Ga, and B are weaker then those of the elements in groups 1A and 2A.

4. Dihydrogen-bonded complexes formed by hydrides of the group 4A elements can be classified as weak or very weak. The dihydrogen bonding energy depends on the nature of the elements and increases in the order C < Si < Ge < Sn. The weakest interactions, C–H· · ·H–C with nonpolarized C–H bonds, should be reformulated as H–H bonding, due to the absence of an electrostatic component in bonding.

## REFERENCES

1. Q. Liu, R. Hoffman, *J. Am. Chem. Soc.* (1995), **117**, 10108.
2. I. Rozas, I. Alkota, J. Elguero, *Chem. Phys. Lett.* (1997), **275**, 423.
3. S. A. Kulkarni, *J. Phys. Chem. A* (1998), **102**, 7704.
4. S. A. Kulkarni, A. K. Srivastava, *J. Phys. Chem. A* (1999), **103**, 2836.
5. I. Alkorta, J. Elguero, C. Foces-Foces, *Chem. Commun.* (1996), 1633.
6. S. J. Grabowski, *J. Phys. Chem. A* (2000), **104**, 5551.
7. I. Alkorta, J. Elguero, O. Mo, M. Yanez, J. E. Del Bene, *J. Phys. Chem. A* (2002), **106**, 9325.
8. S. J. Grabowski, W. A. Sokalski, J. Leszezynki, *J. Phys. Chem. A* (2004), **108**, 5823.
9. I. Alkorta, K. Zborowski, J. Elguero, M. Solimannejad, *J. Phys. Chem.* (2006), **110**, 10279.
10. S. J. Grabowski, *Chem. Phys. Lett.* (1999), **312**, 542.
11. K. N. Robertson, O. Knopp, T. S. Cameron, *Can. J. Chem.* (2003), **81**, 727.
12. A. Hayashi, M. Shiga, M. Tachikawa, *Chem. Phys. Lett.* (2005), **410**, 54.
13. S. J. Grabowski, *J. Mol. Struct.* (2000), **553**, 151.
14. S. J. Grabowski, T. L. Robinson, J. Leszczynski, *Chem. Phys. Lett.* (2004), **386**, 44.
15. M. Solimannejad, I. Alkorta, *Chem. Phys.* (2006), **324**, 459.
16. M. Solimannejad, S. Scheiner, *J. Phys. Chem. A* (2005), **109**, 6137.
17. J. J. Novoa, M. H. Whangbo, J. M. Williams, *J. Chem. Phys.* (1991), **94**, 4835.
18. G. Merino, V. I. Bakhmutov, A. Vela, *J. Phys. Chem. A* (2002), **106**, 8491.
19. T. B. Richardson, S. De Gala, R. H. Crabtree, P. E. M. Siegbahn, *J. Am. Chem. Soc.* (1995), **117**, 12875.
20. O. A. Filippov, A. M. Filin, V. N. Tsupreva, N. V. Belkova, A. Lledos, G. Uiaque, L. M. Epstein, E. S. Shubina, *Inorg. Chem.* (2006), **45**, 3086.
21. Y. Meng, Z. Zhou, C. Duan, B. Wang, Q. Zhong, *J. Mol. Struct. (Theochem)* (2005), **713**, 135.
22. M. Solimannejad, A. Boutalib, *Chem. Phys.* (2006), **320**, 275.
23. O. Gunyadin-Sen, R. Achey, N. S. Datal, A. Stowe, T. Autrey, *J. Phys. Chem. B* (2007), **111**, 677.
24. N. Belkova, O. A. Filippov, A. M. Filin, L. N. Telitskaya, Y. Smirnova, *Eur. J. Inorg. Chem.* (2004), 3453.
25. T. Kar, S. Scheiner, *J. Chem. Phys.* (2003), **119**, 1473.
26. P. C. Singh, G. N. Patwari, *Chem. Phys. Lett.* (2006), **419**, 265.
27. C. J. Cramer, W. L. Gladfelter, *Inorg. Chem.* (1997), **36**, 5358.
28. L. M. Epstein, E. S. Shubina, E. V. Bakhmutova, L. N. Saitkulova, V. I. Bakhmutov, A. L. Chstyakov, I. S. Stankevich, *Inorg. Chem.* (1998), **37**, 3013.
29. E. S. Shubina, E. V. Bakhmutova, A. M. Filin, I. B. Sivaev, V. I. Bakhmutov, A. L. Chstyakov, I. S. Stankevich V. I. Bregadze, L. M. Epstein, *J. Organomet. Chem.* (2002), **657**, 155.
30. O. A. Filippov, A. M. Filin, N. V. Belkova, V. N. Tsupreva, Y. V. Smirnova, I. B. Sivaev, L. M. Epstein, E. S. Shubina, *J. Mol. Struct.* (2006), **790**, 114.
31. T. Steiner, G. R. Desiraju, *Chem. Commun.* (1998), 891.

32. A. J. Barnes, *J. Mol. Struct.* (1983), **100**, 259.

33. A. C. Legon, B. P. Roberts, A. L. Wallrock, *Chem. Phys. Lett.* (1990), **173**, 107.

34. M. T. Nguyen, B. Coussens, L. G. Vanquickenborn, S. Gerber, H. Huber, *Chem. Phys. Lett.* (1990), **167**, 227.

35. M. J. Atkins, A. C. Legon, A. I. Wallrock, *Chem. Phys. Lett.* (1992), **192**, 368.

36. S. L. Benet, F. H. Field, *J. Am. Chem. Soc.* (1972), **94**, 5188.

37. M. G. Govender, T. A. Ford, *J. Mol. Struct. (Theochem)* (2003), **630**, 11.

38. W. L. Zhu, C. M. Puah, X. J. Tan, H. L. Jiang, K. X. Chen, *J. Phys. Chem. A* (2001), **105**, 426.

39. A. H. Pakiari, A. Mohajeri, *J. Mol. Struct. (Theochem)* (2003), **620**, 31.

40. T. B. Richardson, S. Gala, R. H. Crabtree, P. E. M. Siegbahn, *J. Am. Chem. Soc.* (1995), **117**, 12875.

41. C. F. Matta, J. Hernandez-Trujillo, T. H. Tang, R. F. W. Bader, *Chem. Eur. J.* (2003), **9**, 1940.

42. M. Patterson, J. Lundell, M. Rasanen, *Eur. J. Inorg. Chem.* (1999), 729.

43. J. Lundell, M. Patterson, *Phys. Chem. Chem. Phys.* (1999), **1**, 1691.

44. J. Lundell, S. Berski, Z. Latajka, *Phys. Chem. Chem. Phys.* (2000), **2**, 5521.

45. S. Berski J. Lundell, Z. Latajka, *J. Mol. Struct.* (2000), **552**, 223.

# 7

# INTERMOLECULAR DIHYDROGEN BONDING IN TRANSITION METAL HYDRIDE COMPLEXES

Hydrogen atoms bonded to transition metal atoms can show acidic or basic behavior as a function of the ligand environment of the metal atoms [1]. For example, the complexes $HCo(CO)_4$ and $H_2Fe(CO)_4$ are well known to be quite strong acids in solution, particularly in water solution. In contrast, the W–H bond, for example in basic complex trans-$WH(CMes)(dmpe)_2$, shows a surprisingly high ionicity value, close to that in the LiH molecule. Generally, however, according to numerous theoretical and experimental data, hydrogen ligands in transition metal hydrides are negatively charged. Moreover, even the complex $HCo(CO)_4$ acts as a hydride rather than a proton donor in the industrially important hydroformylation reaction.

A partial negative charge on the hydride ligands of transition metal hydride complexes is a prerequisite for dihydrogen bonding. On the other hand, metal atoms in these complexes are electron-rich and at the M–H distances are relatively small and measured between 1.6 and 1.8 Å in Scheme 7.1. Both of the factors create a situation when a proton or a positively charged hydrogen atom in a proton donor with a size of $\leq 1.2$ Å attacks the two closely located targets, one of which is significantly larger. In other words, the targets can be in competition. Moreover, a priori, it is difficult to rule out participation of the d-electrons in H$\cdots$H bonding even in the proton attack on a hydride ligand. That is why the idea of H$\cdots$H bonding was initially an object of criticism. DFT calculations [2] of the model $CpReH(CO)(NO)$ complex, **1**, with three potential proton-accepting centers, Re, H, and NO, provide a good illustration of this simple statement. According to the calculations of **1** in the presence of $H_2O$, all the centers are actually capable of proton accepting to give dihydrogen-bonded complex **2** or **4**, a nonclassical Re$\cdots$H bond in **5**, and a classical hydrogen bond in **3** (Figure 7.1). It is important that these adducts show different geometries. However, their energies are very similar. In other words, the potential energy surface of this and other

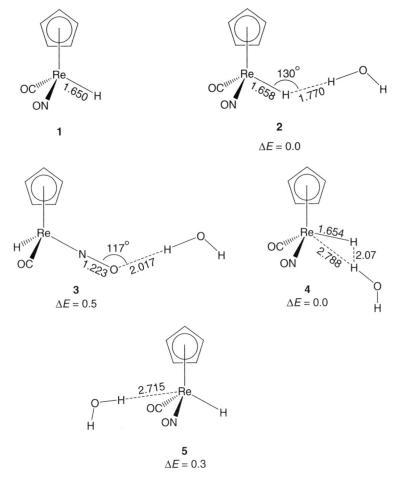

**Scheme 7.1**   Proton attack directed at a metal or hydride center (schematically).

**Figure 7.1**   Geometries of the model Re hydride and the products of its proton attack by a water molecule optimized at the DFT/B3PW91 level with a standard LANL2DZ contraction. Energies of the hydrogen-bonded complexes are given in kcal/mol.

systems can be quite flat, significantly complicating their experimental studies. Nevertheless, Figure 7.1 shows clearly that a dihydrogen bond is still formed as simple or bifurcated, with slight elongations of the initial Re−H bond.

In 1972, Donnel and co-workers suggested that proton–hydride interactions explain transition-state stabilization in the base-catalyzed alcoholysis of silicon triorganohydrides [3]. Later, these interactions have been used to rationalize H/D exchanges and $H_2$ elimination from anionic hydrides $HM(CO)_4L^-$ [4]. In 1992, the attack of a hydride atom by a proton has been considered as an elementary step in kinetics at protonation of transition metal hydrides to yield $H_2$ complexes [5]. Similarly, M−H$\cdots$H−X interaction has been suggested as a factor dictating the fast and highly regioselective proton–hydride exchange shown in Structure 7.1 [6]. Finally, in 1994, Lough and co-workers reported the first experimental evidence for short M−H$\cdots$H−X contacts based on $^1$H $T_1$ NMR relaxation measurements [7]. Later, the idea of dihydrogen bonding became generally accepted and modern experimental and theoretical approaches allow us not only to detect dihydrogen bonds M−H$\cdots$H−X but also to distinguish them from other weak interactions.

Methodological aspects of such studies seem very important and therefore are illustrated here by rhenium dihydrides $ReH_2(CO)(NO)(PR_3)_2$ having five potential proton-acceptor sites: two different Re−H groups to form dihydrogen bonds, groups CO and NO to yield classical hydrogen bonds, and the Re atom to give an unconventional bond Re$\cdots$H.

The existence of an intermolecular dihydrogen bond in the solid state follows clearly from the single-crystal x-ray diffraction structure of adduct $ReH_2(CO)(NO)$ $(PMe_3)_2\cdot$indole, shown in Figure 7.2. One of the hydride atoms located trans to the CO group, shows practically linear geometry of the H$\cdots$H−N fragment and a short ReH(2)$\cdots$H(2$n$) distance of 1.79(5) Å. The latter is similar to the complex indole$\cdots$ReH$_5$(PPh$_3$)$_3$ discussed in Chapter 4. This H$\cdots$H distance is remarkably smaller than the sum of van der Waals radii of H (2.4 Å), thus being a strong argument for dihydrogen bonding. In addition, the free Re−H bond remains short (1.63 Å) while the coordinated Re−H bond is strongly elongated to 2.36 Å. This is also typical of dihydrogen bonding.

In full accord with the structural data, the IR spectrum recorded for the solid complex exhibits two intense bands in the $\nu$(ReH) region assigned to the free and coordinated ReH bonds (Figure 7.3). It is worth mentioning that in the absence of accurate structural data, such $\nu$(Me−H) frequency shifts can serve as a most

**Structure 7.1**

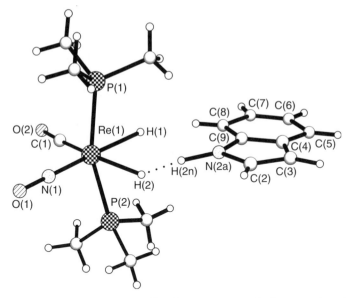

**Figure 7.2**  Single-crystal x-ray diffraction structure of the complex $ReH_2(CO)$ $(NO)(PMe_3)_2$ with indole. (Reproduced with permission from ref. 8.)

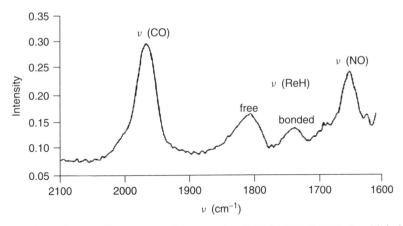

**Figure 7.3**  Solid-state IR spectrum of the complex $ReH_2(CO)(NO)(PMe_3)_2$ with indole. (Reproduced with permission from ref. 9.)

important experimental criterion for recognition of the hydride atom accepting a proton.

The situation becomes more complicated in $CH_2Cl_2$ solutions. For example, the $\nu(OH)$ regions in the IR spectra, recorded for 1 : 1 mixtures of $ReH_2(CO)(NO)$ $(PR_3)_2$ and $(CF_3)_3COH$, show the red frequency $\nu(OH)$ shifts of 243 to 370 $cm^{-1}$. It is obvious that the proton donor participates in intermolecular hydrogen bonding [9]. In accordance with this conclusion, the IR spectra show the new CO bands that appear only in the presence of the proton donor. However, these bands are *high-frequency shifted* by 11 cm$^{-1}$, and thus the CO groups are not involved in hydrogen bond interactions according to the well-known IR spectral criteria. At the same time, the proton donor causes the appearance of two new (NO) bands shifted to higher and lower frequencies, which follows from Figure 7.4. It is seen that intensities of these bands are temperature dependent.

Under these circumstances, the first band may well be assigned to the NO group, which remains free, while the second band belongs to the hydrogen bond OH$\cdots$ON. However, the $\nu$(ReH) region also indicates remarkable changes in the presence of a proton donor. Figure 7.4 shows clearly a low-frequency Re–H shoulder. Again, the intensity of this line increases upon cooling. Thus, in addition to bond OH$\cdots$ON, the Re dihydride forms the dihydrogen bond Re–H$\cdots$HOC(CF$_3$)$_3$. Unfortunately, the IR spectra do not allow us to identify what hydrogen ligand of the dihydride is dihydrogen-bonded. However, this problem can be solved successfully using variable-temperature $^1$H NMR spectra and $^1$H $T_{1,\mathrm{min}}$ NMR measurements in solution [10]. According to these NMR experiments, the hydride atom located trans to the NO group is dihydrogen bonded with an H$\cdots$H distance of 1.78 Å. It is interesting that this distance is very similar to that in the x-ray structure of the indole complex (see Figure 7.2), where the proton donor is binding to the hydrogen atom, which is trans-located to the CO group. Finally, no experimental data demonstrating Re$\cdots$H interaction have been found by IR and NMR experiments.

The results of DFT calculations performed for the model system ReH$_2$(PH$_3$)$_2$(NO)(CO)·H$_2$O in a 1 : 1 ratio have revealed intermolecular interactions with all

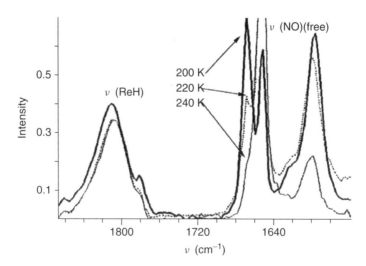

**Figure 7.4** Variable-temperature IR spectra recorded in a CH$_2$Cl$_2$ solution of ReH$_2$(CO)(NO)(PEt$_3$)$_2$ in the presence of PFTB [(CF$_3$)$_3$COH] in the range of $\nu$(NO) and $\nu$(ReH). (Reproduced with permission from ref. 9.)

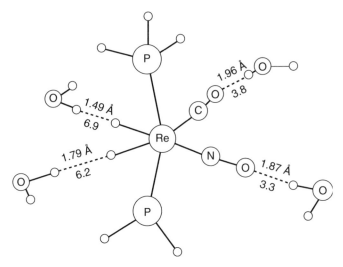

**Figure 7.5**

of the possible proton-acceptor centers, excluding Re· · ·H bonding [9]. It follows
from Figure 7.5 that both hydridic hydrogen atoms accept the proton donor to
form dihydrogen bonds with H· · ·H distances of 1.49 and 1.79 Å, corresponding
to the trans and cis hydrogen locations with respect to the NO group, respectively.
However, the former gives the largest energy of formation, calculated as −6.9
kcal/mol. Then the energy decreases in the order

$$6.2 \text{ kcal/mol (the cis H ligand)} > 3.8 \text{ kcal/mol (bond HOH} \cdots \text{ON–Re)}$$

$$> 3.3 \text{ kcal/mol (bond HOH} \cdots \text{OC–Re)} \quad (7.1)$$

It is seen that the DFT calculations closely reproduce the tendencies observed
experimentally.

Previously, we noted that a dihydrogen bond can be bifurcated or three-centered
(Structure 7.2), with one hydrogen atom tightly bonded while the second shows
weak interaction. Such bonding can be found, for example, in the neutron diffraction structure of the dihydrogen-bonded complex indole· · ·ReH$_5$(PPh$_3$)$_3$ (see

**Structure 7.2**

Figure 4.5). The $ReH_2(PH_3)_2(NO)(CO) \cdot H_2O$ system has been optimized in the framework of this model to give distances of 1.63 and 1.87 Å [9]. However, this structure has resulted in a notably smaller energy gain of $-2.6$ kcal/mol with respect to linear $H \cdots H$ bonds.

To conclude this section, it should be emphasized again that the potential energy surface of transition metal hydrides, interacting with proton donors, is expected to be flat, the type of hydrogen bonding depending on the competition between numerous electronic and steric effects. Under this circumstance, the best way to investigate is through a combination of experimental approaches and theoretical calculations, which probably provide the most valuable results.

## 7.1. THEORETICAL VIEW OF INTERMOLECULAR DIHYDROGEN BONDING IN TRANSITION METAL HYDRIDE COMPLEXES

In contrast with the relatively simple dihydrogen-bonded systems formed by hydrides of the elements in groups 1A to 4A considered previously, accurate calculations of transition metal hydride complexes containing large and electron-rich transition metal atoms represent a well-known computational problem [11]. Among different theoretical approaches, the modified coupled pair functional (MCPF) and coupled cluster [(CCSD(T)] methods are probably most reliable. However, because of the high cost of such computations and the size limitation for the systems calculated, they are not in regular use. On the other hand, good accuracy in calculations of transition metal complexes can be reached by density functional theory (DFT) methods. They usually provide good agreement with experimental data as well as with the results obtained by MP2 calculations.

To determine the best scheme to use for dihydrogen bonding in transition metal hydride complexes, Orlova and Scheiner [11] carried out calculations of the model Mo and W hydride complexes acting as proton acceptors in the presence of a strong acid, HF. The work was performed at the 3-21G, B3LYP, BLYP, and B3PW91 levels. The B and B3 functionals have been used to model the exchange effect, while the correlation effect has been included via the functional LYP and PW91. Figure 7.6 illustrates the data obtained for the Mo hydride, where, in principle, two structures correspond to dihydrogen bonding. In Figure 7.6(a), the H–F molecule is located in the same plane as the Mo, P, and H atoms, whereas it lies outside the plane in the structure on the right. The former complex shows a negative frequency at all theoretical levels, however, and thus represents a transition state. The structure shown in Figure 7.6(b) occupies a minimum area on the potential energy surface. The dihydrogen bonding energy of the transition state has been calculated as $-14.7, -12.4, -12.0$, and $-10.5$ kcal/mol at the 3-21G, BLYP, B3LYP, and B3PW91 levels, respectively, while the ground-state structure on the right was calculated at $-17.6, -12.9, -12.4$, and $-11.1$ kcal/mol at the same levels. Thus, it is clear that the ab initio 3-21G method overestimates remarkably the energy of the dihydrogen bonding.

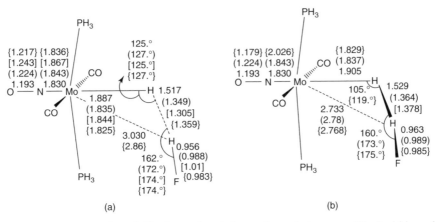

**Figure 7.6** Geometries of dihydrogen-bonded complexes between an Mo hydride and an HF molecule optimized at the 3-21G, B3LYP (parentheses), BLYP (brackets), and B3PW91 (braces) levels. (Reproduced with permission from ref. 11.)

As expected, the geometry, and the H···H distance particularly, are very sensitive to the computing method used. For example, the H···H distance changes within 1.529 and 1.378 Å, where the B3PW91 approach provides the best results. It is interesting, however, that the methods reveal similar structural tendencies in dihydrogen bond formation: The M–H and H–F bonds are remarkably elongated with respect to initial molecules (their optimized geometries are shown in Figure 7.7); the Mo–H···H bond is bent at an angle of $120°$ while the H···H–F fragment is practically linear. We have already observed the same tendencies in solid-state x-ray and neutron diffraction structures. Finally, the calculations reveal insignificant effects on going from Mo to W. For example, the W dihydrogen-bonded complex with geometry close to that of the structure shown in Figure 7.6(b) is formed with an energy gain of $-11.4$ kcal/mol (at the B3PW91 level), which is larger by only 0.3 kcal/mol relative to the Mo complex.

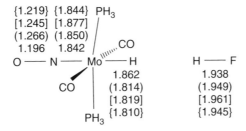

**Figure 7.7** Geometries of an Mo hydride and an HF molecule optimized at the 3-21G, B3LYP (parentheses), BLYP (brackets) and B3PW91 (braces) levels. (Reproduced with permission from ref. 11.)

**TABLE 7.1. Theoretical Energies of Dihydrogen Bonding and H···H Distances in Some Transition Metal Hydride Systems**

| Number | Complex | $-\Delta E^a$ (kcal/mol) | $r(\text{H} \cdots \text{H})$ (Å) | Method |
|---|---|---|---|---|
| 1 | CpNbH$_3$···HOCH$_3{}^b$ | 6.7 (4.1) | 1.792 | B3LYP |
| 2 | CpNbH$_3$···HOOCCH$_3{}^b$ | 8.6 (5.8) | 1.707 | |
| 3 | CpNbH$_3$···HOCH$_2$CF$_3{}^b$ | 8.7 (5.9) | 1.646 | |
| 4 | CpNbH$_3$···HOCH(CF$_3$)$_2{}^b$ | 9.5 (4.5) | 1.635 | |
| 5 | CpNbH$_3$···HOC(CF$_3$)$_3{}^b$ | 11.1 (6.4) | 1.589 | |
| 6 | CpNbH$_3$···HOOCCF$_3{}^b$ | 11.3 (8.7) | 1.587 | |
| 7 | CpNbH$_3$···HOCF$_3{}^b$ | 12.3 (8.0) | 1.670 | |
| 8 | (PH$_3$)$_2$Mo(CO)$_2$(NO)H···HF | 11.1 | 1.378 | B3PW91 |
| 9 | (NH$_3$)$_2$Mo(CO)$_2$(NO)H···HF | 15.4 | 1.300 | |
| 10 | (NH$_3$)$_2$Mo(CO)$_2$(NO)H···HOH | 13.1 | 1.647 | |
| 11 | CpRe(CO)(NO)H···HOH | 10.9 | 1.770 | |
| 12 | CpRe(CO)(PH$_3$)H···HOCF$_3$ | 9.8 | 1.458 | |
| 13 | CpRe(dhpe)H···HOCH$_2$CF$_3$ | 10.46 (5.83) | 1.736 | B3PW91 |
| 14 | CpFe(dhpe)H···HOCH$_2$CF$_3$ | (4.92) | | |
| 15 | CpRu(CO)(PH$_3$)H···HOOCCF$_3$ | (8.4)$^c$ | 1.697 | B3LYP |
| 16 | CpRu(CO)(PH$_3$)H···HOC(CF$_3$)$_3$ | (9.7)$^c$ | 1.654 | |
| 17 | CpMo(dpe)H$_2$H···HOCH(CF$_3$)$_2$ | 9.07 | 1.649 | B3LYP |
| 18 | CpW(dpe)H$_2$H···HOCH(CF$_3$)$_2$ | 9.00 | 1.706 | |
| 19 | PP$_3$RuHH$^{\text{ax}}$···HOCH$_3$ | 9.7 | 1.785 | RHF/LANL2DZ |
| 20 | PP$_3$RuHH$^{\text{eq}}$···HOCH$_3$ | 9.68 | 1.703 | |
| 21 | PP$_3$RuHH$^{\text{ax}}$···HOCF$_3$ | 20.77 | 1.395 | |
| 22 | PP$_3$RuHH$^{\text{eq}}$···HOCF$_3$ | 17.35 | 1.335 | |

*Source:* Data from refs. 11 to 17.

$^a$BSSE-corrected values are shown in parentheses.

$^b$Calculated for the central hydride ligand.

$^c$5.6 and 2.4 kcal/mol are calculated for this dihydrogen-bonded complex by computer simulation of the solvent polarity for *n*-heptane and CH$_2$Cl$_2$.

Table 7.1 lists energies of dihydrogen bonding and the H···H distances that have been calculated for transition metal hydride systems in the gas phase. As shown, the transition metal complexes calculated have different ligand environments and interact with different proton donors.

The niobium trihydride (1 to 7) illustrates clearly the influence of proton-donor strength: The dihydrogen bonding energy increases from the weak proton donor CH$_3$OH ($-6.7$ kcal/mol), to the relatively strong acid CF$_3$OH ($-12.3$ kcal/mol).

This energy effect agrees well with shortening the H$\cdots$H distances. Quantitatively close results are observed for the Ru dihydride on going from 19 to 21 or from 20 to 22. The effects of ligand environments also seem to be significant. For example, comparing cases 8 and 9 shows that an increase in cis-$\sigma$ donor strength of the ligand from PH$_3$ to NH$_3$ is accompanied by increasing the energy gain by 4.3 kcal/mol. At the same time, the influence of the metal atom is minimal. As we show below, these tendencies are easily reproduced experimentally.

However, it is important to note that the energy values, calculated for gas-phase dihydrogen-bonded complexes are usually overestimated with respect to the experimental measurements carried out in solutions. The latter is connected directly with solvent effects. In fact, the computer simulation of solvent polarity demonstrates significant lowering of the energy of formation: for example, from 6.7 to 5.2 and 2.4 kcal/mol, computed for niobium trihydride system 1 on going from the gas phase to $n$-heptane and CH$_2$Cl$_2$.

## 7.2. ENERGY AND STRUCTURAL PARAMETERS OF INTERMOLECULAR DIHYDROGEN-BONDED COMPLEXES IN SOLUTIONS OF TRANSITION METAL HYDRIDES

The formation of dihydrogen-bonded complexes in solution can be established independently by two spectroscopic methods: IR spectroscopy and $^1$H NMR. According to IR data, the bands of hydrogen-bond acceptor sites, $\nu$(MH), can undergo the lower (or higher)-frequency shifts while the bands of hydrogen-bond donors, $\nu$(OH), shift to lower frequencies. The $^1$H NMR data show that dihydrogen bonding leads to *high-field shifts* of MH resonances and the $^1$H $T_1$ times of these resonances decrease due to additional proton–hydride dipole–dipole interactions with protons of HX. Finally, accurate evaluations of the dipole–dipole contributions provide accurate determinations of the H$\cdots$H distances. At the same time, since the world of transition metal hydrides is quite diverse and the spectral behavior of dihydrogen-bonded systems is different, it is difficult to formulate a uniform recipe for their quantitative characterizations. For example, the $^1$H $T_1$ time treatments for the Nb trihydride complexes are complicated due to the presence of large H–Nb dipole–dipole interactions [18]. In addition, trihydride CpNbH$_3$ is a single representative of early transition metal hydrides forming dihydrogen bonds [12]. Therefore, it seems to be reasonable to illustrate some details that are important to identification of such dihydrogen-bonded systems.

Trihydride CpNbH$_3$ shows two resonances in the hydride region of variable-temperature $^1$H NMR spectra, corresponding to the central and lateral hydride ligands. Chemical shifts of these resonances are practically independent of temperature. In contrast, in the presence of (CF$_3$)$_2$CHOH, both resonances undergo high-field shifts that increase on cooling (Figure 7.8). On the other hand, as indicated in the spectra, the central resonance moves remarkably more strongly, thus demonstrating stronger dihydrogen bonding. $^1$H NMR relaxation measurements performed at 400 MHz for both resonances demonstrate the appearance of additional dipole–dipole contributions coming from the proton of (CF$_3$)$_2$CHOH: $^1$H

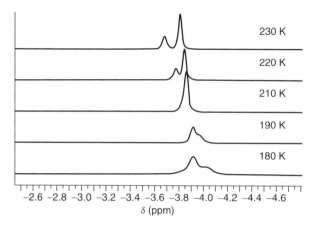

**Figure 7.8** Variable-temperature $^1$H NMR spectra of CpNbH$_3$ in the presence of (CF$_3$)$_2$CHOH (1 : 1) in toluene-d$_8$. The less intense line belongs to the central hydride ligand. (Reproduced with permission from ref. 12.)

$T_{1,\text{min}} = 0.086$ and $0.106$ s for the central and lateral hydride ligands, respectively, versus $0.109$ and $0.132$ s in the absence of (CF$_3$)$_2$CHOH. However, in contrast to many cases considered earlier in the book, simple comparison of relaxation rates in the absence and presence of the proton donor does not lead to accurate evaluation of the dipole–dipole H$\cdots$H contributions because of very pronounced dipole interactions Nb–H. This evaluation requires variable-temperature $T_1$ measurements performed for a partially deuterated system that contains a mixture of isotopomers CpNbDH$_2$, CpNbD$_2$H, and (CF$_3$)$_2$CHOD. According to these measurements in combination with the $T_1$ data collected for CpNbH$_3$ [12], the H$\cdots$H distances can be determined as 1.92 and 1.84 Å for the lateral and central hydride ligands, respectively. Again, the central hydride shows stronger dihydrogen bonding. Thus, despite the additional difficulties, $^1$H $T_1$ relaxation remains a powerful technique for structural investigations in solutions. Some of the H$\cdots$H distances are given in Table 7.2.

Based on numerous experimental studies, the set of the dihydrogen-bonding energies obtained for different transition metal hydride complexes is very representative (Table 7.2). It should be noted that most energy parameters has been obtained via $\nu$(OH) shifts and the corresponding $\Delta\nu/\Delta E$ relationships. Some of them, such as $\Delta H^\circ$, have been measured directly by variable-temperature IR spectra (see numbers 5, 11, 16 to 18, 26, and 36). It is worth noting that in these cases the entropy values were significant and negative, varying between $-9.8$ and $-23.8$ eu. This is in good agreement with the bimolecular character of dihydrogen bonding.

The data in Table 7.2 show that the dihydrogen-bonding energy increases with the proton-donor strength of proton donors. In this context, systems 28 to 31 are most representative where the $\Delta E$ value changes from $-4.1$ kcal/mol for CH$_3$OH to $-6.6$ kcal/mol for (CF$_3$)$_2$CHOH. In accordance with the theoretical data, the

**TABLE 7.2. Experimental Energies, H· · ·H Distances, Enthalpy ($\Delta H°$), and Entropy ($\Delta S°$) Values (eu) Determined for Transition Metal Dihydrogen-Bonded Complexes in Solution**

| Number | Complex | $-\Delta E^a$ /r(HcdotsH)$^b$ [(kcal/mol)/Å] | Solvent | Ref |
|--------|---------|-----------|---------|-----|
| 1 | CpNbH$_3$· · ·HOCH$_2$CF$_3$$^c$ | 4.6/1.92$^d$ | CH$_2$Cl$_2$ | 12 |
| 2 | CpNbH$_3$· · ·HOCH(CF$_3$)$_2$$^c$ | 5.7/1.84$^d$ | CH$_2$Cl$_2$ | 12 |
| 3 | PP$_3$FeH$_2$· · ·HOCH$_3$ | 3.3 | CH$_2$Cl$_2$ | 17 |
| 4 | PP$_3$FeH$_2$· · ·HOCH$_2$CH$_2$F | 3.8 | CH$_2$Cl$_2$ | 17 |
| 5 | PP$_3$FeH$_2$· · ·HOCH$_2$CF$_3$ | 4.5 | CH$_2$Cl$_2$ | 17$^e$ ($-\Delta H° = 5.1$; $-\Delta S° = 13.4$) |
| 6 | PP$_3$FeH$_2$· · ·HOCH(CH$_3$)$_2$ | 3.6 | CH$_2$Cl$_2$ | 17 |
| 7 | PP$_3$RuH$_2$· · ·HOCH$_2$CF$_3$ | 5.4 | CH$_2$Cl$_2$ | 17 |
| 8 | PP$_3$RuH$_2$· · ·HOCH(CF$_3$)$_2$ | 6.4 | CH$_2$Cl$_2$ | 17 |
| 9 | PP$_3$RuH$_2$· · ·HOCH$_3$ | 4.8 | CH$_2$Cl$_2$ | 17 |
| 10 | PP$_3$OsH$_2$· · · HOCH$_2$CH$_2$F | 5.8 | CH$_2$Cl$_2$ | 17 |
| 11 | PP$_3$OsH$_2$· · ·HOCH$_2$CF$_3$ | 6.7/1.96$^d$ | CH$_2$Cl$_2$ | 17$^e$ ($-\Delta H° = 6.8$; $-\Delta S° = 19.2$ |
| 12 | Cp*Mo(dppe)H$_3$ · · ·HOCH$_2$CH$_2$F | 4.9 | CH$_2$Cl$_2$ | 16 |
| 13 | Cp*Mo(dppe)H$_3$ · · ·HOCH$_2$CF$_3$ | 5.9 | CH$_2$Cl$_2$ | 16 |
| 14 | Cp*W(dppe)H$_3$ · · ·HOCH$_2$CH$_2$F | 6.0 | CH$_2$Cl$_2$ | 16 |
| 15 | Cp*W(dppe)H$_3$ · · ·HOCH$_2$CF$_3$ | 6.7 | CH$_2$Cl$_2$ | 16 |
| 16 | CpRu(CO)(PCy$_3$)H · · ·HOC(CF$_3$)$_3$ | | Hexane | 15$^e$ ($-\Delta H° = 7.3$; $-\Delta S° = 21.6$) |
| 17 | Cp*Ru (dppe)H · · ·HOCH$_2$CH$_2$F | 4.7 | CH$_2$Cl$_2$ | 14$^e$ ($-\Delta H° = 4.5$; $-\Delta S° = 15.8$) |
| 18 | Cp*Ru (dppe)H · · ·HOCH$_2$CF$_3$ | 5.7/1.40$^d$ | CH$_2$Cl$_2$ | 14$^e$ ($-\Delta H° = 5.7$; $-\Delta S° = 23.8$) |
| 19 | (PMe$_3$)$_2$Re(NO)(CO)H$_2$ · · ·HOC(CF$_3$)$_3$ | 6.1/1.78$^d$ | Hexane | 9 |
| 20 | (PEt$_3$)$_2$Re(NO)(CO)H$_2$ · · ·HOC(CF$_3$)$_3$ | 5.2 | Hexane | 9 |

(*continued*)

**TABLE 7.2.** (*Continued*)

| Number | Complex | $-\Delta E^a$ /$r$(HcdotsH)$^b$ [(kcal/mol)/Å] | Solvent | Ref |
|--------|---------|-----------------------------------------------|---------|-----|
| 21 | $(PPr_3)_2Re(NO)(CO)H_2$ $\cdots HOC(CF_3)_3$ | 4.5 | Hexane | 9 |
| 22 | $CpRu(CO)(PCy_3)H$ $\cdots HOCCH_2CF_3$ | 5.3 | Hexane | 19 |
| 23 | $CpRu(CO)(PCy_3)H$ $\cdots HOCH(CF_3)_2$ | 6.3 | Hexane | 19 |
| 24 | $CpRu(CO)(PCy_3)H$ $\cdots HOCH(CF_3)_2$ | 5.1 | $CH_2Cl_2$ | 19 |
| 25 | $CpRu(CO)(PCy_3)H$ $\cdots HOC(CF_3)_3$ | 7.6 | Hexane | 19 |
| 26 | $Cp*Fe(dppe)H$ $\cdots HOCH_2CH_2F$ | 4.6 | $CH_2Cl_2$ | 20$^e$ ($-\Delta H° = 5.4$; $-\Delta S° = 13.6$ |
| 27 | $Cp*Fe(dppe)H$ $\cdots HOCH_2CF_3$ | 5.9 | $CH_2Cl_2$ | 20 |
| 28 | $(PP_3)Ru(CO)(H)H$ $\cdots HOCH_3$ | 4.1 | $CH_2Cl_2$ | 21 |
| 29 | $(PP_3)Ru(CO)(H)H$ $\cdots HOCH_2CH_2F$ | 4.7 | $CH_2Cl_2$ | 21 |
| 30 | $(PP_3)Ru(CO)(H)H$ $\cdots HOCH_2CF_3$ | 5.8 | $CH_2Cl_2$ | 21 |
| 31 | $(PP_3)Ru(CO)(H)H$ $\cdots HOCH(CF_3)_2$ | 6.6/1.81$^d$ | $CH_2Cl_2$ | 21 |
| 32 | $(NP_3)H_3Re\cdots HOCH_2CH_3$ | 4.5 | $CH_2Cl_2$ | 22 |
| 33 | $(NP_3)H_3Re\cdots HOCH_2CH_2F$ | 4.9 | $CH_2Cl_2$ | 22 |
| 34 | $(NP_3)H_3Re\cdots HOCH_2CF_3$ | 6.3 | $CH_2Cl_2$ | 22 23 |
| 35 | $(PP_3)Re(CO)_2H$ $\cdots HOC(CF_3)_3$ | 6.0/1.83$^d$ | $CH_2Cl_2$ | 24$^e$ ($-\Delta H° = 4.9$; $-\Delta S° = 9.8$) |
| 36 | $(PEt_3)_2W(CO)_2(NO)H$ $\cdots HOCH(CF_3)_2$ | 5.1 | Hexane | |

$^a$Determined by IR $\nu$(OH) spectra via the Iogansen equation.
$^b$Determined by the $^1H$ $T_{1,min}$ NMR relaxation experiments.
$^c$Determined for the central hydride ligand.
$^d$In toluene-$d_8$.
$^e$Determined by variable-temperature IR spectra.

influence of the ligand environment is significant (see, e.g., the cis effects in systems 19 to 21). Finally, the solvent polarity lowers the energy gain, as is clear from a comparison of the data obtained in hexane and $CH_2Cl_2$ for systems 23 and 24.

Dihydrogen-bonded complexes 5, 7, and 11 demonstrate the influence of the metal atom. Moving along the row of the periodic table leads to increases in energy gain: −4.5 (Fe), −5.4 (Ru), and −6.7 (Os) kcal/mol. The same order is observed for systems 12 and 14: $\Delta E = -4.9$ kcal/mol for Mo and −6.0 kcal/mol for W. It is interesting that the differences in energy, caused by the nature of the metal, are significant and more pronounced with respect to those predicated theoretically. It is probable that this is connected to the influence of solvents.

The data in Table 7.2 show that the H···H distances are determined in solutions between 1.40 and 1.96 Å, strongly supporting formulation of the interactions as dihydrogen bonds. Finally, systems 32 to 34 in Table 7.2 are shown for comparison, and it is clear that dihydrogen bonds and nonclassical M···H bonds are energetically very similar.

## 7.3. UNUSUAL M–H···H–M BONDS

Since the electrostatic component in C–H···H–C contacts is completely absent, Bader has termed these very weak interactions H–H interactions or H–H bonding but not dihydrogen bonds (see Chapter 6). The second example of such a surprising interaction can be found in the crystal packing of monohydride β-MH(CO)$_5$ reported by Calhorda and Costa [25]. According to x-ray and neutron diffraction data, the structure of this hydride is associated with the presence of dimers, where an intermolecular H···H distance is determined as 2.292 Å (i.e., smaller than the van der Waals sum; see Figure 7.9). Due to the presence of the powerful electron-accepting groups CO, the hydride ligand in this molecule is positively charged. However, again, this situation corresponds to the absence of any electrostatic attraction between these hydrogen atoms.

To rationalize the nature of this H···H interaction, the authors have performed HF, DFT, and MP2 calculations to optimize the geometry for the monomer and dimer molecules as well. The extended Hückel approach has resulted in the molecular orbital diagram shown in Figure 7.10, according to which the weak H···H contact is a four-electron destabilizing interaction between the σ M–H bonding levels, where mixing of the σ* M–H levels still produces a weakly bonding character. In some sense, this delicate electron balance is similar to that suggested for $H_2$ binding to a transition metal in dihydrogen complexes. However, in this case the H–H moiety is binding to two metal atoms.

DFT/B3LYP calculations have resulted in an H···H distance of 2.008 Å, which is very close to the experimental data. On the other hand, the bonding energy is negligibly small in this case (∼1 kcal/mol). Calculations at the HF/6-311G** level did not localize the dimer species at an energy minimum of the potential energy surface, while the MP2/6-311G** level has shown that this dimer truly occupies

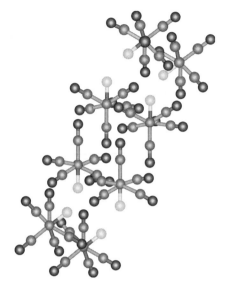

**Figure 7.9** Crystal packing in the β-form of monohydride MnH(CO)₅, illustrating the relatively short H···H contacts. (Reproduced with permission from ref. 25.)

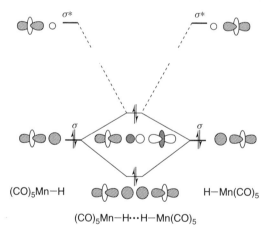

**Figure 7.10** Molecular orbital diagram obtained for the dimer [(Mn(CO)₅]₂ in the framework of extended Hückel calculations. (Reproduced with permission from ref. 25.)

an energy minimum. The latter illustrates an important role for correlation effects that contribute to the energy of the dimer. The energy gain for the [MnH(CO)₅]₂ was 10.7 kcal/mol higher than twice the energy of isolated monomers. The value transforms to −5.1 kcal/mol after BSSE correction. However, in this case the H···H distance is too short (1.51 Å). This effect could be connected with the well-known overestimation of weak interactions in the MP2 framework. Finally, the authors emphasize that calculations at the various levels allow us to think that this $H\cdots H$ interaction is stronger than a classical $H\cdots O$ interaction.

## 7.4. CONCLUDING REMARKS

1. Dihydrogen bonds of transition metal hydrides are of medium strength or even strong in the gas phase (the theoretical data) since their bonding energy changes between 6.7 and 15.4 kcal/mol at the $H\cdots H$ distance 1.3 to 1.8 Å. In solution, they are closer to medium strength, with a typical energy between 3 and 8 kcal/mol. The typical $H\cdots H$ distance value is lies between 1.6 and 1.9 Å. When the strength of proton donors increases, the dihydrogen bonding energy increases and the $H\cdots H$ distance decreases.

2. Dihydrogen bonding depends on the nature of the metal; moving along the row leads to increased energy gain.

3. Unusual $M-H\cdots H-M$ interaction is rationalized in the following terms: The weak $H\cdots H$ contact is a four-electron destabilizing interaction between the $\sigma$ $M-H$ bonding levels, where mixing of the $\sigma^*$ $M-H$ levels produces a weakly bonding character.

## REFERENCES

1. H. Jacobsen, H. Berke, in *Recent Advances in Hydride Chemistry*, ed. M. Peruzzini and R. Poli, Elsevier, New York (2001), pp. 89–116.
2. G. Orlova, S. Scheiner, *J. Phys. Chem. A* (1998), **102**, 4813.
3. K. O. Donnel, R. Bacon, K. L. Chellapa, R. L. Schowen, J. K. Lee, *J. Am. Chem. Soc.* (1972), **94**, 2500.
4. P. L. Gaus, S. C. Kao, M. Y. Darensbourg, L. W. Arndt, *J. Am. Chem. Soc.* (1984), **106**, 4752.
5. P. G. Jessop, R. H. Morris, *Coord. Chem. Rev.* (1992), **121**, 155.
6. S. Feracin, T. Burgi, V. I. Bakhmutov, I. Eremenko, E. V. Vorontsov, A. B. Vimenits, H. Berke, *Organometallics* (1994), **13**, 4194.
7. A. J. Lough, S. Park, R. Ramachandaran, R. H. Morris, *J. Am. Chem. Soc.* (1994), **116**, 8356.
8. N. V. Belkova, E. S. Shibina, E. I. Gutsul, L. M. Epstein, S. E. Nefedov, I. L. Eremanko, *J. Organomet. Chem.* (2000), **610**, 58.
9. N. V. Belkova, E. S. Shubina, A. V. Ionidis, L. M. Epstein, H. Jacobson, A. Messmer, H. Berke, *Inorg. Chem.* (1997), **36**, 1522.
10. A. Messmer, H. Jacobsen, H. Berke, *Chem. Eur. J.* (1999), **5**, 3341.
11. G. Orlova, S. Scheiner, *J. Phys. Chem. A* (1998), **102**, 260.
12. E. V. Bakhmutova, V. I. Bakhmutov, N. V. Belkova, M. Besora, L. M. Epstein, A. Lledos, G. V. Nikonov, E. S. Shubina, J. Tomas, E. V. Vorontsov, *Chem. Eur. J.* (2004), **10**, 661.
13. G. Orlova, S. Scheiner, T. Kar, *J. Phys. Chem. A* (1999), **103**, 514.

14. N. V. Belkova, P. A. Dub, M. Baya, J. Houghton, *Inorg. Chim. Acta* (2007), **360**, 149.

15. N. V. Belkova, M. Besora, L. M. Epstein, A. Lledos, F. Maseras, E. S. Shubina, *J. Am. Chem. Soc.* (2003), **125**, 7715.

16. J. Andrieu, N. V. Belkova, M. Besora, E. Collange, L. M. Epstein, A. Lledos, R. Poli, P. O. Ravin, E. S. Shubina, E. V. Vorontsov, *Russ. Chem. Bull.* (2003), **52**, 2679.

17. E. I. Gutsul, N. V. Belkova, M. S. Sverdlov, L. M. Epstein, E. S. Shubina, V. I. Bakhmutov, T. N. Gribanova, R. M. Minaev, C. Bianchini, M. Peruzzini, F. Zanobini, *Chem. Eur. J.* (2003), **9**, 2219.

18. V. I. Bakhmutov, *Practical Nuclear Magnetic Resonance Relaxation for Chemists*, Wiley, Chichester, England (2004).

19. N. V. Belkova, A. V. Ionidis, L. M. Epstein, E. S. Shubina, S. Gruendemann, N. S. Golubev, H. H. Limbach, *Eur. J. Inorg. Chem.* (2001), 1753.

20. N. V. Belkova, P. O. Revin, L. M. Epstein, E. V. Vorontsov, V. I. Bakhmutov, E. S. Shubina, E. Collange, R. Poli, *J. Am. Chem. Soc.* (2003), **125**, 11106.

21. V. I. Bakhmutov, E. V. Bakhmutova, N. V. Belkova, C. Binachini, L. M. Epstein, D. Masi, M. Peruzzini, E. S. Shubina, E. V. Vorontsov, F. Zanobini, *Can. J. Chem.* (2001), **79**, 479.

22. A. Albinati, V. I. Bakhmutov, N. V. Belkova, C. Binachini, I. Rios, L. M. Epstein, E. I. Gutsul, L. Marvelli, M. Peruzzini, R. Rossi, E. S. Shubina, E. V. Vorontsov, F. Zanobini, *Eur. J. Inorg. Chem.* (2002), 1530.

23. N. V. Belkova, E. V. Bakhmutova, E. S. Shubina, C. Binachini, M. Peruzzini, V. I. Bakhmutov, L. M. Epstein, *Eur. J. Inorg. Chem.* (2000), 2165.

24. E. S. Shubina, N. V. Belkova, A. N. Krylov, E. V. Vorontsov, L. M. Epstein, D. G. Gusev, M. Niedermann, H. Berke, *J. Am. Chem. Soc.* (1996), **118**, 1105.

25. M. J. Calhorda, P. J. Costa, *Cryst. Eng. Commun.* (2002), **4**, 368.

# 8

# CORRELATION RELATIONSHIPS FOR INTERMOLECULAR DIHYDROGEN BONDS

A correlation analysis is a powerful tool used widely in various fields of theoretical and experimental chemistry. Generally, such an analysis, based on a statistically representative mass of data, can lead to reliable relationships that allow us to predict or to estimate important characteristics of still unknown molecular systems or systems unstable for direct experimental measurements. First, this statement concerns structural, thermodynamic, kinetic, and spectroscopic properties. For example, despite the very complex nature of chemical screening in NMR, particularly for heavy nuclei, various incremental schemes accurately predict their chemical shifts, thus providing a structural analysis of new molecular systems. Relationships for the prediction of physical or chemical properties of compounds or even their physiological activity are also well known.

When correlation relationships imply a deep physical context, they become a good basis for the development of new, more common concepts or for the systematization of compounds, their properties, reactions, and even phenomena. Such relationships characterizing dihydrogen bonds are the focus of the chapter.

## 8.1. GENERAL CLASSIFICATION OF NEGATIVELY POLARIZED HYDROGEN ATOMS AS PROTON-ACCEPTING SITES: BASICITY FACTORS

As we have shown, dihydrogen and classical hydrogen bonds are similar geometrically and energetically despite the fact that in the former case a hydrogen bond acceptor and a hydrogen bond donor are hydrogen atoms. A hydrogen atom accepting a proton is basic, and therefore quantitative characterization of the basicity is of great interest. This is particularly important for transition metal

*Dihydrogen Bonds: Principles, Experiments, and Applications*, By Vladimir I. Bakhmutov
Copyright © 2008 John Wiley & Sons, Inc.

hydride complexes, which are key components in catalysis, where hydride migration to a small molecule bound to a metal center is a typical step in catalytic cycles involved, for example, in hydrogenation, hydrocyanation, or olefin polymerization [1].

Hydride basicity can be systemized by studies of migration-insertion behavior of transition metal hydrides [2] or determinations of deuterium quadrupole coupling constants for deuterium ligands that are a measure of $M-H$ bond ionicity [3]. Another approach to systematization can be based on treatment of dihydrogen-bonding energies, measured experimentally or found theoretically. This approach is close to the *rule of factors* suggested by Iogansen in 1981 [4] in an attempt to generalize the numerous experimental data collected for classical hydrogen bonds.

This simple but very effective idea is based on an analysis of the large mass of calorimetric and IR spectroscopic data and directed to a quantitative comparison of the proton-accepting ability of various groups that form classical hydrogen bonds. The rule suggests constants $P_i$ and $E_j$ as measures of the acidic and basic strengths, respectively. These constants are related as follows:

$$E_j = \frac{\Delta H_{ij}^\circ}{\Delta H_{11}^\circ P_i} \tag{8.1}$$

where $-\Delta H_{ij}^\circ$ is the formation energy, determined experimentally for a hydrogen-bonded complex and $-\Delta H_{11}^\circ$ is the value taken as a standard. The latter characterizes the hydrogen bond formed by the phenol (as a proton donor) and diethyl ether (as a proton acceptor). Then the acidity and basicity factors for this standard pair should be taken as $P_i = E_j = 1$. It is worth mentioning that $E_j$ and $P_i$ in eq. (8.1) are generalized and therefore are completely independent of the nature of the proton donor and the solvent.

Table 8.1 shows the $P_i$ factors for some acidic components capable of hydrogen bonding, which as might be expected, increase from $CH_3OH$ to $(CF_3)_2COH$. It should be noted that this order of acidity was used by us intuitively in earlier

**TABLE 8.1. Acidity Factors Scaling Some Proton-Donor Molecules**

| Acid | Acidity Factor $P_i$ |
|------|----------------------|
| $Pr^iOH$ | 0.58 |
| $CH_3OH$ | 0.63 |
| $FCH_2CH_2OH$ | 0.74 |
| Indole | 0.75 |
| $CF_3CH_2OH$ | 0.89 |
| $(CF_3)_2CHOH$ | 1.06 |
| $(CF_3)_3COH$ | 1.35 |

*Source:* Ref. 5.

discussions of dihydrogen bonds. Since hydrogen and dihydrogen bonds are similar and electrostatic contributions dominate in both cases, the same approach has been employed for quantitative characterizations of hydridic hydrogens participating in dihydrogen bonding. The energies of this bonding have been determined by IR spectra. As expected, the energy of dihydrogen bonding correlates with the $P_i$ factor. For example, a nice linear relationship has been observed for dihydrogen-bonded complexes, where hydridic hydrogens in $[BH_4]^-$, $Et_3NBH_3$, and $P(OEt_3)BH_3$ play the role of proton-accepting sites (Figure 8.1).

The basicity factors, $E_j$, obtained for various boron and transition metal hydrides are collected in Table 8.2. As shown, the $E_j$ values vary from 0.41 in boron hydride, $P(OEt)_3BH_3$, to 1.66 in the dihydride $PP_3OsH_2$. The former compound represents a weak base due to the electron-attractive $P(OEt)_3$ group. In contrast, the hydride $PP_3OsH_2$ has an $E_j$ value that is even larger than that measured for dimethyl sulfoxide (DMSO) (1.27). Moreover, the basicity of the hydrogen ligands in $PP_3OsH_2$ is close to that of pyridine (1.67), which is known as quite a strong organic base.

Since the $E_j$ factors have been obtained from the energies of dihydrogen bonding, they reproduce accurately the tendencies connected with changes in the nature of the metal and its ligand environment. For example, the $E_j$ factor as a function of the ligand environment increases from $(EtO)_3PBH_3$ to $Et_3NBH_3$ and $[B_{12}H_{12}]^-$. It is interesting that the basicity of hydrogen atoms in the ions $[BH_4]^-$ and $[GaH_4]^-$ is similar to that of the oxygen electron pair in DMSO. Then, for

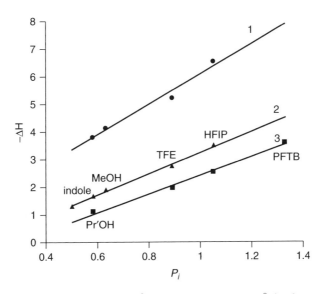

**Figure 8.1** Linear dependencies $\Delta H^\circ$ versus $P_i$, where $\Delta H^\circ$ is the energy of the dihydrogen-bonded complexes (kcal/mol) and $P_i$ is the acidity factor obtained for $[BH_4]^-$ in (1) $CH_2Cl_2$, (2) $Et_3NBH_3$ and (3) $P(OEt_3)BH_3$ in hexane. PFTB is $(CF_3)_3COH$, TFE is $CF_3CH_2OH$, and HFIP is $(CF_3)_2CHOH$. (Reproduced with permission from ref. 6.)

**TABLE 8.2. Basicity Factors Characterizing Hydridic Hydrogens in Boron and Transition Metal Hydrides**

| Proton-Accepting Site | Basicity Factor $E_j$ |
|---|---|
| $P(OEt)_3BH_3$ | 0.41 |
| $Et_3NBH_3$ | 0.53 |
| $ReH_7(dppe)$ | 0.54 |
| $[B_{12}H_{12}]^{2-}$ | 0.61 |
| $ReH_5(PPh_3)_3$ | 0.63 |
| $WH(CO)_2(NO)(PPh_3)_2$ | 0.70 |
| $WH(CO)_2(NO)(POPr^i_3)_2$ | 0.72 |
| $[B_{10}H_{10}]^{2-}$ | 0.78 |
| $ReH_2(CO)(NO)(PMe_3)_2$ | 0.80 |
| $BH_3CN^-$ | 0.80 |
| $WH(CO)_2(NO)(PEt_3)_2$ | 0.87 |
| $WH(CO)_2(NO)(PMe_3)_2$ | 0.91 |
| $[MeC(CH_2PPh_2)_3]Re(CO)_2H$ | 0.97 |
| $CpRuH(CO)PCy_3$ | 1.0 |
| $PP_3FeH_2$ | 1.0 |
| $[BH_4]^-$ | 1.25 |
| $PP_3RuH_2$ | 1.33 |
| $[GaH_4]^-$ | 1.37 |
| $[MeC(CH_2PPh_2)_3]Ru(CO)H_2$ | 1.39 |
| $Ru(dppm)_2H_2$ | 1.40 |
| $PP_3OsH_2$ | 1.66 |
| $Et_2O$ | 1.00 |
| DMSO | 1.27 |
| Py | 1.67 |

*Source*: Data from refs. 7 to 9.

transition metal hydride complexes having identical ligand environments, the $E_j$ value increases in the order

$$1.0 \text{ (Fe)} < 1.33 \text{ (Ru)} < 1.6 \text{ (Os)} \tag{8.2}$$

as a function of the nature of the metal.

Finally, it should be emphasized that the $E_j$ factor can be a good indicator, showing the electron density distribution in *the ground state* of proton acceptors (i.e., hydridic hydrogens) when dihydrogen bonding induces small changes on theses centers: in other words, when only electrostatic interaction contributes to the $\Delta H°$ (i.e., $E_j$) value. In reality, even relatively simple dihydrogen-bonded systems such as $Li-H\cdots H-X$ or $Na-H\cdots H-X$ exhibit remarkable polarization and charge transfer contributions. This circumstance will probably complicate use of the $E_j$ parameters. For example, charged dihydrogen-bonded complexes such as that in Structure 8.1 show a tendency that is opposite to eq. (8.2). Here, the dihydrogen bond strength reduces down the group $Ru > Os$ [10].

**Structure 8.1**

## 8.2. CORRELATION RELATIONSHIPS: ENERGY OF FORMATION VERSUS H···H DISTANCE ESTABLISHED FOR INTERMOLECULAR DIHYDROGEN BONDS

Correlations of the type described in the heading have a fundamental significance since they form a basis for discussions of dihydrogen bonds in terms that are widely applicable for covalent chemical bonds. In other words, the existence of such correlations allows us to consider dihydrogen bonding as a part of chemical bonds.

The relationships connecting bonding energies and distances between a proton and a proton-accepting center are well known for medium strength and strong classical hydrogen bonds such as O–H···O. However, as has been noted, there is no rationalization of these relationships other than an intuitive one [11]. In the context of the energy–distance relationships, the following aspect is very important: Classical hydrogen bonds, O–H···O, can be very stable thermodynamically, with bonding energies up to 20 to 30 kcal/mol, but highly dynamic at the same time. This situation is typical of the enol–enolic systems depicted in Scheme 2.3, where short O···O distances of 2.35 to 2.5 Å correspond to low-barrier hydrogen bonds. It is worth mentioning that the height of the potential barrier for proton transfer can be as low as 1 kcal/mol (sometimes the proton in such systems is centered between two oxygen atoms). These systems illustrate an important principle: Proton transfer along a classical hydrogen bond occurs faster when this hydrogen bond is stronger. As we show in Chapter 10, the same principle works for dihydrogen bonding in transition metal hydrides, where fast proton transfer to hydrogen ligands leads to the formation of dihydrogen complexes or $H_2$ elimination. For such short-living dihydrogen bonds, correlation relationships are particularly important.

Figure 8.2 shows a good linear correlation between the $-\Delta H^\circ$ values and the H···H distances found for dihydrogen-bonded complexes formed by $[AlH_4]^-$, $[BH_4]^-$, $[GaH_4]^-$, and Cp*Fe(dhpe)H with different proton donors [5,7]. In this case the bonding energy increases from 5 kcal/mol to 12 kcal/mol and the H···H distance changes between 1.3 and 1.7 Å. It is probable that the linearity of this relationship is connected with the relatively narrow diapasons mentioned above. In fact, larger diapasons for both parameters lead to the polynomial dependence [12] shown in Figure 8.3. This pattern includes the dihydrogen-bonded complexes

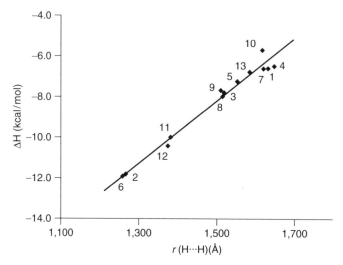

**Figure 8.2** Enthalpies of dihydrogen-bonded complexes versus H···H distances. The dihydrogen-bonded complexes are formed between different proton-donor components [$CH_3OH$, $CF_3OH$, $CH_3CH_2OH$, $FCH_2CH_2OH$, $(CF_3)_3COH$, and $CF_3COOH$] and [$AlH_4$]$^-$ (points 1 to 3), [$BH_4$]$^-$ (points 4 and 5) [$GaH_4$]$^-$ (points 6 to 8), and Cp*Fe(dhpe)H (points 9 to 13) as proton-accepting sites. (Reproduced with permission from ref. 7.)

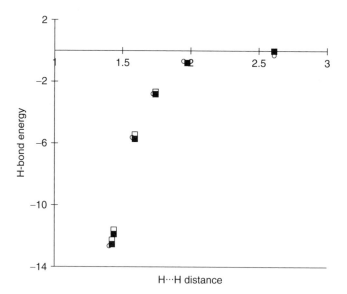

**Figure 8.3** Relationships between the energy of formation (kcal/mol) and the H···H distances (Å) calculated for dihydrogen-bonded complexes of HF and LiH, NaH, $MgH_2$, $BeH_2$, $Mo(CO)_2(NO)(PH_3)_2H$, $CH_4$, and $SiH_4$ as proton-acceptor sites: circles, the MP4/6-311++G** level; open squares, the QCISD/6-311++G** level; solid squares, the QCISDT/6-311++G** level. (Reproduced with permission from ref. 12.)

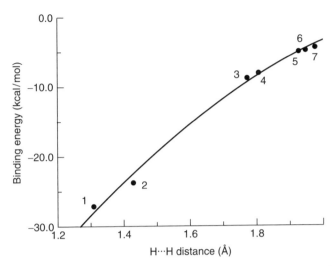

**Figure 8.4** Energies of formation versus H$\cdots$H distances calculated for C–H$\cdots$H-Li dihydrogen-bonded complexes: 1, LiNCH$^+\cdots$HLi; 2, NaNCH$^+\cdots$HLi; 3, NCH$\cdots$HLi; 4, NCCCH$\cdots$HLi; 5, ClCCH$\cdots$HLi; 6, FCCH$\cdots$HLi; and 7, HCCH$\cdots$HLi. (Reproduced with permission from ref. 13.)

formed by an Mo hydride and hydrides of group 1A, 2A, and 4A elements: LiH, NaH, MgH$_2$, BeH$_2$, CH$_4$, and SiH$_4$. In this case the bonding energy changes from $-13$ kcal/mol to 0 kcal/mol and the H$\cdots$H distance increases from 1.3 Å to 2.6 Å. It should be emphasized that the shape of the curve changes insignificantly with the theoretical level.

A similar nonlinear dependence (Figure 8.4) has been reported by Alkorta and co-workers [13]. The authors calculated a set of dihydrogen-bonded complexes where LiH plays the role of a strong proton acceptor while the C–H bond represents a proton-donor component whose strength changes within wide limits. The analytic expression of this dependence is

$$BE = -19.3r^2 + 99.37r - 124.85 \qquad (8.3)$$

It is interesting that calculations of dihydrogen-bonded complexes CH$_4\cdots$HF, SiH$_4\cdots$HF, BeH$_2\cdots$HF, LiH$\cdots$HF, NaH$\cdots$HF, GeH$_4\cdots$HF, SnH$_4\cdots$HF and CrH$_2\cdots$HF [14] show a very similar shape of dependence, expressed analytically as

$$BE = -35.3r^2 + 136.18r - 131.55 \qquad (8.4)$$

Thus, all the data support the idea that the polynomial shape is most common.

The same set of dihydrogen-bonded complexes has been probed to establish the relationship between the H$\cdots$H and H–F distances in Figure 8.5. This dependence deserves a more detailed discussion. In fact, as we will see below, it can

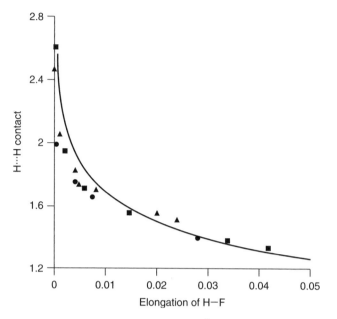

**Figure 8.5** Relationship between H$\cdots$H distances (Å) and elongations of the H–F bond (Å) calculated for dihydrogen-bonded complexes CH$_4\cdots$HF, SiH$_4\cdots$HF, BeH$_2\cdots$HF, LiH$\cdots$HF, NaH$\cdots$HF, GeH$_4\cdots$HF, SnH$_4\cdots$HF, and CrH$_2\cdots$HF. (Reproduced with permission from ref. 14.)

be interpreted in the frameworks of the bond valence model [15]. According to this model, the strength of the regular chemical bond or of the intermolecular contact can be measured by bond valences. The relationship between the bond valance and the bond length can be approximated as

$$s_{ij} = \exp\left(\frac{r_0 - r_{ij}}{B}\right) \quad \text{or} \quad s_{ij} = \left(\frac{r_{ij}}{r_0}\right)^{-N} \tag{8.5}$$

where $s_{ij}$ is the bond valence of the bond between atoms $i$ and $j$, $r_{ij}$ is the bond length, and $B$ and $N$ are well-known constants [16]. In these terms, $r_0$ is the length of a bond of unit valence; that is the length of the single bond is not deformed by hydrogen bonds or by any other intermolecular interaction. According to the bond valence model, atomic valences $V_i$ (equal to the oxidation states of atoms) are assumed to be shared between the bonds they form:

$$\sum s_{ij} = V_i \tag{8.6}$$

Then, in the context of dihydrogen bonds, for example, X–H$\cdots$H–F, the valence rule can be written as

$$s(\text{H} - \text{F}) + s(\text{H} \cdots \text{H}) = 1 \tag{8.7}$$

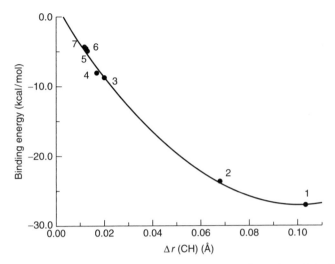

**Figure 8.6** Relationship between the binding energies and elongation of the C–H bonds in the C–H···H–Li dihydrogen-bonded complexes (for a key to the compound numbers, see Figure 8.4). (Reproduced with permission from ref. 13.)

Thus, this simple equation implies the relation between the H···H distance and the H–F bond length, which is shown in Figure 8.5. Here squares represent the data obtained at the MP2/6-311++G* and triangles show the MP2/6-31G** data. The solid line is obtained via eqs. 8.7 and 8.5 using the $B$ and $r_0$ constants given by Dunitz [16]. This line reflects the validity of the bond valence model.

Since the bonding energy correlates with the H···H distance, which in turn is connected with the elongation of the H–X bonds in proton donors, it is reasonable to expect the relationships between the bonding energies and H–X bond elongation. Figure 8.6 illustrates one such dependency, obtained by Alkorta and co-workers for the dihydrogen-bonded complexes Li–H···H–C [13], where the larger bonding energy corresponds to greater elongation of the C–H bonds in proton donors.

It is obvious that in addition to the theoretical importance of the results presented, which provide a better understanding of dihydrogen bonds, these relationships can be used for a rough estimation of the H···H distances if the bonding energy has already been determined by a convenient method. Finally, readers are referred to refs. 17 to 21 for details of calculations and additional results.

## 8.3. CORRELATION RELATIONSHIPS AMONG ENERGETIC, STRUCTURAL, AND ELECTRON DENSITY PARAMETERS OF INTERMOLECULAR DIHYDROGEN-BONDED COMPLEXES

As mentioned previously, AIM theory suggests a more accurate formulation of dihydrogen bonding, based on the topology of the electron density analyzed in

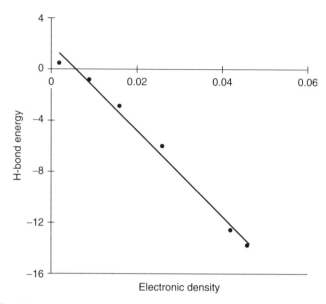

**Figure 8.7** Relationship between the electronic density at the critical point in the H$\cdots$H direction (in electronic units) and the energy of formation calculated for LiH$\cdots$HF, NaH$\cdots$H–F, BeH$_2\cdots$H–F, MgH$_2\cdots$HF, CH$_4\cdots$H–F and SiH$_4\cdots$H–F at the MP2/6-311++G** level. (Reproduced with permission from ref. 12.)

the H$\cdots$H directions. In this connection, it is very interesting to trace the relationships between the dihydrogen bonding energies or H$\cdots$H distances and the electron density parameters calculated for bond critical points, such as $\rho_C$ and $\nabla^2\rho_C$. Grabowskii [12] found the linear dependence energy versus $\rho_C$ shown in Figure 8.7, obtained by treatment of dihydrogen-bonded complexes between HF and the hydrides LiH, NaH, BeH$_2$, MgH$_2$, CH$_4$, and SiH$_4$. Since this dependence involves the large set of elements in groups 1A, 2A, 3A, and 4A, it may be a useful tool for rough determinations of dihydrogen bond energies when the electron density at bond critical points is known, or vice versa. It is interesting that such dependence is again linear for dihydrogen complexes C–H$\cdots$H–Li calculated at a large variation in the proton-donor strength of the C–H component (Figure 8.8).

Robertson and co-workers [18] carefully analyzed a very big number of dihydrogen bonds, from very strong, such as in the H$_2$ molecule or Li–H$\cdots$H–F, to very weak, such as C–H$\cdots$H–C. The authors reported that the difference between the Mulliken charges at the hydrogen atoms in individual proton donors and proton acceptors correlates linearly with the parameters in dihydrogen-bonded complexes. It is obvious that this correlation will be very useful for predictions of the topological strength for the H$\cdots$H bonds. However, among numerous results obtained in this work, analysis of electron density as a function of the H$\cdots$H distance is of greatest interest. It follows from Figure 8.9 that the electron density, $\rho_A$, increases monotonically with reduced H$\cdots$H distance, $d_A$. However, the function is not linear and can be described as an exponential dependence. This

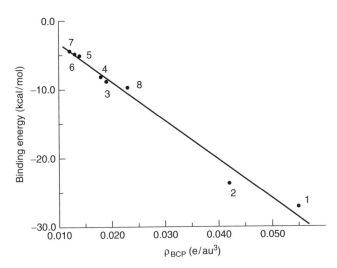

**Figure 8.8** Energies of formation versus the $\rho_C$ parameter determined for bond critical points localized in the H$\cdots$H directions in C–H$\cdots$H–Li dihydrogen-bonded complexes: 1, LiNCH$^+\cdots$HLi; 2, NaNCH$^+\cdots$HLi; 3, NCH$\cdots$HLi; 4, NCCCH$\cdots$HLi; 5, ClCCH$\cdots$HLi; 6, FCCH$\cdots$HLi; and 7, HCCH$\cdots$HLi. (Reproduced with permission from ref. 13.)

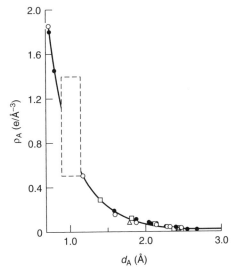

**Figure 8.9** Electron density $\rho$ at bond critical points found in the H$\cdots$H directions versus the H$\cdots$H distance in an $H_2$ molecule and dihydrogen-bonded complexes of Li–H, $BeH_2$, $BH_3$, $BH_4^-$, LiCCH, and $CH_4$ with various proton-donor components: HF, $HNF_3^+$, $HNH_3^+$, HNC, HCN, $HCCSiH_3$, HCCH, HCCF, $HCCCH_3$, HCCLi, and $CH_4$. The "taboo" domain between 0.9 and 1.15 Å, shown as the rectangle, corresponds to the case when the dihydrogen-bonded complexes are not stable enough to yield a free $H_2$ molecule, or are nonexistent. (Reproduced with permission from ref. 18.)

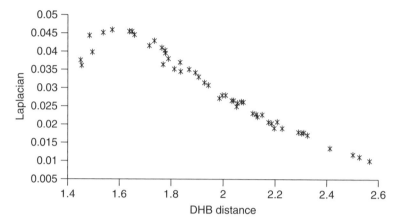

**Figure 8.10** Dependence of the Laplacian values, $\nabla^2 \rho_C$ (au), on the H$\cdots$H distances (Å) calculated for dihydrogen-bonded complexes formed by the hydrides LiH, NaH, BeH$_2$, and MgH$_2$ with the proton donors HCN, HNC, and HCCH, including monomer–dimer and dimer–dimer combinations of proton donors and proton acceptors. (Reproduced with permission from ref. 17.)

result agrees well with theoretical studies of $^1$H–$^1$H spin–spin coupling constants through bonds H$\cdots$H (see, e.g., Figure 3.12). The Fermi contact contribution to this spin–spin coupling [see the $\delta$-function in eq. (3.7)] can be considered as a measure of the electron density on the H$\cdots$H direction. As follows from Figure 3.12, this Fermi contact constant decreases monotonically with increasing H$\cdots$H distance and again nonlinearly.

Evolution of the Laplacian, $\nabla^2 \rho_C$, as an important criterion for dihydrogen bonding with H$\cdots$H distance has been traced by Alkorta and co-workers. [17] The nonmonotonic character of the curve, shown in Figure 8.10, is not surprising. In fact, in the framework of AIM theory, the maximum, located at a distance of about 1.6 Å, precludes the negative Laplacian values expected at small atom–atom distances when interaction becomes open shell: in other words, the bond transforms to covalent. Finally, the authors emphasize that excluding this region yields a good exponential function that describes the $\nabla^2 \rho_C$/distance dependence. Similar results are described in ref. 18.

## 8.4. PROTON AFFINITY OF HYDRIDIC HYDROGENS

The concept of proton affinity is well known to play a very important role in organic chemistry. In fact, this concept permits us to express quantitatively the proton-accepting strength of organic bases. It is significant that the proton affinity arises from the nature of a proton donor.

It is generally accepted that the two most important factors affecting the structure and energies of classical hydrogen bonds are the acidity of donors and the

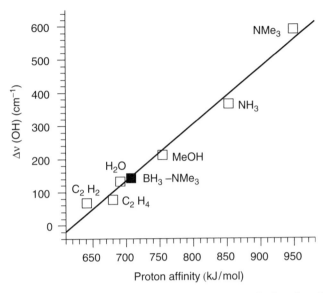

**Figure 8.11** Change in the O–H stretching frequency of phenol calculated for hydrogen-bonded complexes formed by various proton acceptors versus their proton affinities. (Reproduced with permission from ref. 22.)

proton affinity of acceptors. In other words, one can postulate that hydrogen bonding energy is *proportional* to the proton affinity of the acceptor. It is well recognized that lowering the proton-donor stretching frequency under hydrogen bonding is linearly correlated with the proton affinity of the acceptor. Thus, this lowering effect can be used to determine proton affinity on the basis of stretching vibrations of proton donors observed experimentally or calculated theoretically.

The statement above shows that the proton affinity and the basicity factor, $E_j$, are ideologically similar. The latter, determined in solution (Table 8.2), shows that the proton-accepting strength of hydridic hydrogens is close to that of regular organic bases. Patwari and co-workers have determined the proton affinity of borane amines in the gas phase [22,23]. The authors have carried out MP2/6-31G(d) calculations for hydrogen- and dihydrogen-bonded complexes formed by different proton donors. They have been focused particularly on changes in vibrational frequencies $\Delta v(C–H)$, $\Delta v(N–H)$, and $\Delta v(OH)$ obtained for proton donors under complexation. Figure 8.11 illustrates data obtained for phenol acting as a proton donor, where the proton affinity of the B–H hydrogen in $BH_3N(CH_3)_3$ is estimated as 710 kJ/mol (see the solid square). Thus, this proton affinity is higher than that of the oxygen in $H_2O$, but it is lower than that in $CH_3OH$. In accord with the general representations, the proton affinity of borane amines (Figure 8.12) increases in line with the number of $CH_3$ groups. In addition, the same tendency is observed for regular amines when they form classical hydrogen bonds. Thus, these data confirm the idea that hydrogen and dihydrogen bonding are similar.

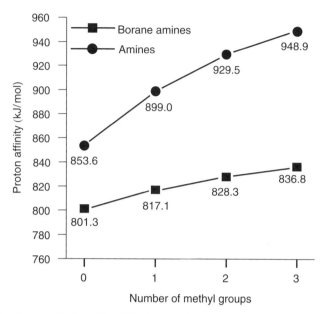

**Figure 8.12** Proton affinity of B–H hydrogens in borane amines and nitrogen atoms in amines versus number of $CH_3$ groups. (Reproduced with permission from ref. 22.)

A more careful comparison of these two types of bonding has been performed for the molecules $CH_3OCH_3$ and $BH_3N(CH_3)_3$ interacting with different linear and nonlinear proton donors [23]. Table 8.3 shows the proton affinities (PAs) where the average PA value obtained for $BH_3N(CH_3)_3$ is remarkably different for linear and nonlinear complexes (695 versus 720 kJ/mol, respectively). This is in contrast with the proton affinity of $CH_3OCH_3$, which forms classical hydrogen bonds. Moreover, as shown in the table, the PA values of $BH_3N(CH_3)_3$ are much less than the values calculated by PA(G2) theory [24]. Thus, the PA values determined for $BH_3N(CH_3)_3$ underestimate the proton affinity of hydridic hydrogen. In other words, these values, found by the method developed for hydrogen bonding, require scaling factors. The latter show that dihydrogen bonding is not

**TABLE 8.3. Proton Affinities (kJ/mol) Characterizing the $CH_3OCH_3$ and $BH_3N(CH_3)_3$ Molecules at Hydrogen and Dihydrogen Bonding to Linear (a) and Nonlinear (b) Proton Donors**

| Acceptor | HCN | $C_2H_2$ | HF | $H_2O$ | $CH_3CONH_2$ | $CH_3COOH$ | PA | PA(G2)[a] |
|---|---|---|---|---|---|---|---|---|
| $CH_3OCH_3$ | 786 | 786 | 776 | 787 | 778 | 776 | 781 | 792 |
| (a) $BH_3N(CH_3)_3$ | 692 | 697 | 697 | — | — | — | 695 | 837 |
| (b) $BH_3N(CH_3)_3$ | — | — | — | 717 | 726 | 718 | 720 | 837 |

[a]PA(G2) is calculated on the basis of the G2 theory [24].

a simple variation of hydrogen bonding. Finally, it is noted that a hydrogen bond is about 16 to 20% stronger than a corresponding dihydrogen bond [23].

## 8.5. CONCLUDING REMARKS

1. The concepts of basicity factor and proton affinity applied to hydridic hydrogens make it possible to express their proton-accepting strength quantitatively, with respect to that of regular organic bases. Both of the factors following from the nature of a proton donor show that the basicity of hydridic hydrogens varies within wide limits, in some cases reaching the basicity level of oxygen atoms in $CH_3OCH_3$ or DMSO.

2. The dependencies "bonding energy/H$\cdots$H distance," "bonding energy/electron density $\rho_C$" "electron density $\rho_C$/H$\cdots$H distance" represent monotonic curves. Generally, they are nonlinear and polynomial in shape. In addition to their theoretical importance, these relationships can be used for a rough estimation of the structural or electronic parameters of dihydrogen if the binding energy has already been determined by a convenient method.

## REFERENCES

1. R. H. Crabtree, *The Organometallic Chemistry of the Transition Metals*, Wiley, Hoboken, NJ (2001).

2. J. A. Labinger, K. M. Komadina, *J. Organomet. Chem.* (1978), **155**, C25.

3. V. I. Bakhmutov, *Practical Nuclear Magnetic Resonance Relaxation for Chemists*, Wiley, Chichester, England (2004).

4. A. V. Iogansen, *The Hydrogen Bond*, Nauka, Moscow (1981), p. 134.

5. N. V. Belkova, E. S. Shubina, L. M. Epstein, *Acc. Chem. Res.* (2005), **38**, 624.

6. L. M. Epstein, E. S. Shubina, E. V. Bakhmutova, L. N. Saitkulova, V. I. Bakhmutov, A. L. Chstyakov, I. S. Stankevich, *Inorg. Chem.* (1998), **37**, 3013.

7. N. Belkova, O. A. Filippov, A. M. Filin, L. N. Telitskaya, Y. Smirnova, *Eur. J. Inorg. Chem.* (2004), 3453.

8. N. V. Belkova, E. I. Gutsul, E. S. Shubina, L. M. Epstein, *Z. Phys. Chem.* (2003), **217**, 1525.

9. L. M. Epstein, E. S. Shubina, *Coord. Chem. Rev.* (2002), **231**, 165.

10. D. G. Gusev, A. J. Lough, R. H. Morris, *J. Am. Chem. Soc.* (1998), **120**, 13138.

11. M. Ichikawa, *Acta Crystallogr.* (1978), **B34**, 2074.

12. S. J. Grabowski, *J. Phys. Chem. A* (2000), **104**, 5551.

13. I. Alkorta, J. Elguero, O. Mo, M. Yanez, J. E. Del Bene, *J. Phys. Chem. A* (2002), **106**, 9325.

14. S. J. Grabowski, *Chem. Phys. Lett.* (1999), **312**, 542.

15. I. D. Brown, *Acta Crystallogr. B* (1992), **48**, 553.

16. J. D. Dunitz, *X-ray Analysis and the Structure of Organic Molecules*, Cornell University Press, London (1979).

17. I. Alkorta, K. Zborowski, J. Elguero, M. Solimannejad, *J. Phys. Chem. A* (2006), **110**, 10279.
18. K. Robertson, O. Knop, T. S. Cameron, *Can. J. Chem.* (2003), **81**, 727.
19. S. J. Grabowski, W. A. Sokalski, J. Leszczynski, *J. Phys. Chem. A* (2004), **108**, 5823.
20. S. J. Grabowski, *J. Mol. Struct.* (2000), **553**, 151.
21. M. G. Govender, T. Ford, *J. Mol. Struct. (Theochem)* (2003), **630**, 11.
22. G. N. Patwari, *J. Phys. Chem. A* (2005), **109**, 2035.
23. P. C. Singh, G. N. Patwari, *J. Phys. Chem. A* (2007), **111**, 3178.
24. L. A. Curtiss, K. Raghavachari, G. W. Trucks, J. A. Pople, *J. Chem. Phys.* (1991), **94**, 7221.

# 9

# PERSPECTIVES ON DIHYDROGEN BONDING IN SUPRAMOLECULAR CHEMISTRY AND CRYSTAL ENGINEERING

Weak noncovalent intra- and intermolecular interactions are centrally important for an understanding of molecular recognition as a fundamental phenomenon in biological processes. In the context of supramolecular chemistry and crystal engineering, directed to the creation of new materials that have unusual physicochemical properties, this phenomenon corresponds to the reproducibility of topological properties in a variety of structural environments. By Desiraju's definition, "crystal engineering is the planning and synthesis of a crystal structure from its molecular constituents; the assembly of the constituents is mediated by intermolecular interactions termed supramolecular glue" [1]. It is obvious that directionality plays a very important role in this glue interaction.

In this context, the classical hydrogen bonds provide the best combination of strength and directionality, permitting rapid self-organization of molecular building blocks into extended regular structures. This process is very efficient with respect to conventional supramolecular synthesis using covalent bonds.

Whereas crystal engineering on the basis of conventional hydrogen bonds is now a well-explored field, only recently have crystal engineers began to explore the potential of nonclassical hydrogen bonds as directors of molecular associations, especially $C-H\cdots X-B$ (X = Cl, Br, I) bonds and $C-H\cdots H-M$ bonds [2]. Despite the quite limited number of systematic studies performed in this field, the strength and directionality of hydridic–protonic interactions, which are comparable to those of conventional hydrogen bonds, indicate that we should anticipate their growing use for crystal packing and molecular associations [3]. It should be added that even weak intramolecular interactions $CH\cdots HC$ can contribute to the stabilization of hydrocarbon molecular crystals, affecting their sublimation energies (see Chapter 6). Here we consider the role of dihydrogen bonding in molecular assemblies and investigate a strategy for its application in crystal engineering.

*Dihydrogen Bonds: Principles, Experiments, and Applications*, By Vladimir I. Bakhmutov
Copyright © 2008 John Wiley & Sons, Inc.

Strategically, there are two perspective ideas suggested for practical applications of dihydrogen bonds. The first is rather traditional and exploits dihydrogen bonding as an interaction that directly connects molecular building blocks. The second idea is based on molecular blocks built via dihydrogen bonds that form a supramolecular system through new covalent bonds appearing due to solid-state $H_2$ elimination:

$$Y-H \cdots H-X \rightarrow Y-X + H_2 \qquad (9.1)$$

Following Custelcean and co-workers, "this great advantage of dihydrogen bonds opens new avenues to the rational assembly of extended covalent materials with controlled architecture" [4]. As examples, we describe in this chapter some derivatives of icosahedral carboranes widely applied as building blocks in supramolecular systems, as well as $NaNH_4 \cdot$ poly(2-hydroxyethyl)cyclen building blocks.

One of the important aspect of the use of dihydrogen bonds as driving forces in molecular associations is their cooperativity or anticooperativity when increasing the number of $H \cdots H$ contacts in the self-association of molecular systems leads to additional energy, calculated on the basis of one dihydrogen bond, or vice versa. As we have shown in Chapter 2, classical hydrogen bonds can be highly cooperative, due to their mutual polarization [5].

The $BH_3NH_3$ molecule can be a good theoretical model for probing the energy effects connected with multidihydrogen bonds. By analogy with hydrogen bond systems [5], the monomeric, dimeric, trimeric, and tetrameric structures of $BH_3NH_3$ have been optimized by high-level DFT calculations [6]. In addition, topological analysis of the electron density in the $H \cdots H$ directions has been performed. Table 9.1 lists the electron density parameters and energies referred to the monomeric structure. Figures 9.1 to 9.3 show the optimized bond lengths, bond paths, and bond critical points obtained in the framework of AIM theory. Frequency analysis has shown that all these structures occupy true energy minima on the potential energy surface. In addition, the geometry of the monomer $NH_3BH_3$ is in good agreement with the gas-phase structure, where the $B-N$ bond is elongated significantly with respect to the solid state. In some sense, the molecular aggregation modeled symbolizes transfer of the $NH_3BH_3$ molecule

**TABLE 9.1. Number of Dihydrogen Contacts ($N$), B–N Bond Length, Electron Density Parameters, and Relative Energies Obtained for Fully Optimized Geometries of $BH_3NH_3$**

| Structure | $N$ | $\rho_C(H-H)$ (au) | $\nabla^2\rho_C(H-H)$ (au) | $r(B-N)$ (Å) | $\rho_C(B-N)$ (au) | $\nabla^2\rho_C(B-N)$ (au) | $\Delta E$ kcal/mol | $\Delta E_{ZPE}$ kcal/mol |
|---|---|---|---|---|---|---|---|---|
| Monomer | 0 | 0 | 0 | 1.665 | 0.100 | 0.406 | 0 | 0 |
| Dimer | 2 | 0.016 | 0.045 | 1.639 | 0.110 | 0.388 | 12.8 | 11.2 |
| Trimer | 4 | 0.014 | 0.040 | 1.620 | 0.118 | 0.370 | 22.7 | 19.9 |
| Tetramer | 6 | 0.012 | 0.036 | 1.607 | 0.125 | 0.362 | 29.7 | 28.5 |

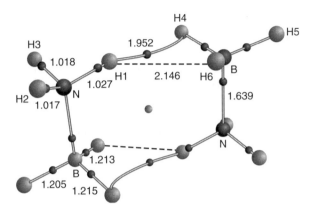

**Figure 9.1** Molecular graph of a dimeric $NH_3BH_3$ molecule with bond and ring critical points. (Reproduced with permission from ref. 6.)

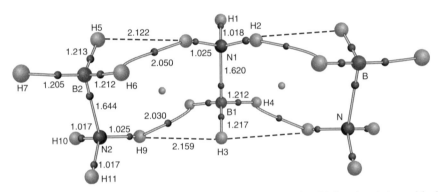

**Figure 9.2** Molecular graph of a trimeric $NH_3BH_3$ molecule with bond and ring critical points. (Reproduced with permission from ref. 6.)

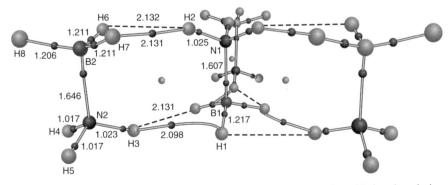

**Figure 9.3** Molecular graph of a tetrameric $NH_3BH_3$ molecule with bond and ring critical points. (Reproduced with permission from ref. 6.)

from the gas phase to the solid state. As we see, in the solid state the B−N bond becomes notably shorter. The tetrameric structure shows the shortest B−N bond.

The DFT data show clearly that the associated $BH_3NH_3$ molecules exhibit a number of relatively short H···H contacts, which seem to be shorter than 2.4 Å. However, only some of them can actually be classified as dihydrogen bonds [compare, e.g., the contacts H(1)···H(4) and H(1)···H(6) in Figure 9.1]. In fact, these dihydrogen bonds show bond critical points with very small $\rho_C(H–H)$ values and positive Laplacians. In contrast, the $\nabla^2\rho_C(B–N)$ values, although positive, are very large, in good agreement with the donor–acceptor character of the central B−N bond.

It follows from topological analysis of the electron density that the number of the truly established dihydrogen bonds increases from two to six from dimer to tetramer. However, with an increasing number of dihydrogen bonds, leading to self-association, the H···H distances *elongate* slightly and the electron density parameters $\rho_C(H–H)$ and $\nabla^2\rho_C(H–H)$ *decrease*. Thus, at least on the level of the structural characteristics and topological parameters, the B−H···H−N dihydrogen bonds do not show a cooperative effect. This conclusion agrees well with the energetic data shown in Table 9.1, where the energy gain, calculated for single H···H contact as 5.6, 4.97 and 4.75 kcal/mol, decreases slightly from dimer to trimer and tetramer, respectively. On the other hand, it is difficult to say that these dihydrogen bonds are anticooperative, as has been shown for some classical hydrogen-bonded systems, where one proton-acceptor group forms two or more hydrogen bonds. Finally, these data demonstrate that even in the absence of cooperative effects, increasing the number of the B−H···H−N dihydrogen bonds leads to a large stabilization energy. This is a very important prerequisite for any organizing interaction in the creation of molecular assemblies.

In the context of self-association processes, the solid-state structures of cyclotriborazane, cyclotrigallazane, and cyclotrialumazane are of great interest [3]. Figure 9.4 illustrates the neutron diffraction structure of cyclotrigallazane, where individual molecules, existing in the chair conformation, bind via Ga−D···D−N dihydrogen bonds with D···D distances of 1.97 Å. These bonds form an α-network parallel to the crystallographic axis. According to RHF and MP2 calculations performed on a cc-pVDZ basis, each H···H contact contributes about 3 kcal/mol to the total formation energy of cyclotriborazane and cyclotrigallazane. This contribution increases slightly in cyclotrialumazane [3].

Since hydrogen and dihydrogen bonds have comparable energies and directionality, their combination could be particularly intriguing to crystal engineers. These bonding interactions have recently been compared by Filippov and co-workers [7], who investigated both theoretically and experimentally the behavior of the ion $[BH_3CN]^-$ in the molecule $[Ph_3P=N=PPh_3]^+[BH_3CN]^-$ in the presence of various OH and NH proton donors in low polar media. Since the $[BH_3CN]^-$ ion has two potential proton-accepting centers, H−N and C−N, the studies have revealed competition between dihydrogen- and hydrogen-bonded complexes. Figure 9.5 shows some of the complexes, structure **3**, which shows dihydrogen and hydrogen bonding simultaneously, being the most interesting. The bonding energies of these complexes are listed in Table 9.2.

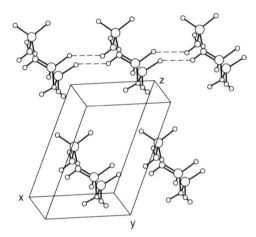

**Figure 9.4** Packing arrangement in one plane of cyclotrigallazane. The dashed lines show the shortest H···H distances. (Reproduced with permission from ref. 3.)

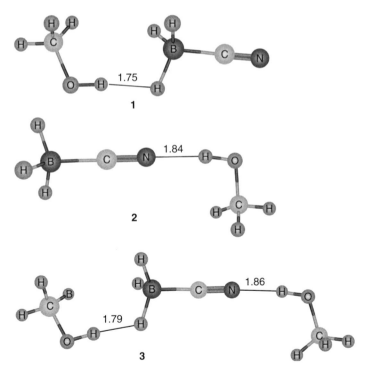

**Figure 9.5** Optimized geometries of the complexes formed by the anion $[BH_3CN]^-$ and methanol. (Reproduced with permission from ref. 7.)

**TABLE 9.2. Energy of Formation for Complexes 1 to 3 of Figure 9.5**[a]

| Complex | $\Delta E$(calc) (kcal/mol) | $\Delta E$(IR) (kcal/mol) |
|---|---|---|
| $[BH_3CN]^- \cdots HOCH_3$ | $-12.09$ | $-6.76$ |
| $[NCBH_3]^- \cdots HOCH_3$ | $-8.23$ | $-4.53$ |
| $[BH_3CN]^- \cdot 2CH_3OH$ | $-19.47$ | $-4.12(N \cdots HO)$ |
| | | $-6.33(BH \cdots HO)$ |

[a]Obtained by DFT(B3LYP)/6-311++G(d,p) calculations and measured experimentally by IR spectra.

According to the data obtained, complex **3** is an example of the coexistence of two bonding types, both of which cause elongation of the H$\cdots$H and N$\cdots$H bonds with respect to 1 : 1 systems. However, the $\Delta E$ value obtained for hydrogen/dihydrogen-bonded complex **3** is close to the a sum of the energies obtained for complexes **1** and **2**. Again, no cooperative effects are observed. At the same time, the total energy gain is significant.

Planas and co-workers, working in carborane supramolecular chemistry, have found that dihydrogen bonding can still provide a cooperative effect [8]. It should be noted that this chemistry applies three-dimensional icosahedral carborane $closo$-$C_2B_{10}H_{12}$ cages as building blocks using the acidic $C_c$–H proton donors ($C_c$ is a cage carbon) to form $C_c$–H$\cdots$X bonds with classical (X = O, N, S, Hal) and weak nonclassical (X = alkynes, arenes) proton acceptors. Multibonding interactions are most interesting in the $closo$-[3-Ru($C_6H_6$)-8-HS-1,2-$C_2B_9H_{10}$] system, shown schematically in Figure 9.6. The extremely low acidity of the boron-attached SH group in this compound is well known, and for this reason it is difficult to expect SH involvement in any significant hydrogen bond interaction. However, the solid-state structure of this system reveals a surprising supramolecular architecture of blocks A, B, and C. It should be added that this trimer of molecules is the unit that is repeated in this two-dimensional polymeric network. Despite the low acidity expected, the SH group still participates in C–H$\cdots$S–H$\cdots$(H–B)$_2$ hydrogen/bifurcated dihydrogen bonding. However, this conclusion follows from the corresponding short contacts. To support the formulation of these interactions as hydrogen/dihydrogen bonds, Planas et al. analyzed this structure theoretically. The geometry of the molecular system was optimized by HF calculations on a LANL2DZ basis for the Ru and S atoms and a 6-311++G(3df,3pf) basis for all the atoms involved directly in the hydrogen–dihydrogen interaction. The optimized minimum of the potential energy surface gave a reasonable reproduction of the experimental results. In good agreement with AIM criteria, the SH$\cdots$H contacts calculated were actually shorter than 2.4 Å, and the geometry showed deviation of the S–H$\cdots$H angle from linearity due to interaction of the SH group with two hydride atoms from a boron cage. The H$\cdots$H–B angles (around 100°) were also typical of dihydrogen-bonded complexes. Finally, in agreement with the bifurcated nature of the complex, Figure 9.7 shows two bond critical points with typical electron density parameters.

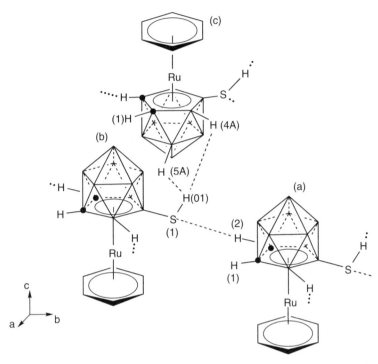

**Figure 9.6** Trimer structure repeated in the two-dimensional network of *closo*-[3-Ru(C$_6$H$_6$)-8-HS-1,2-C$_2$B$_9$H$_{10}$]. (Reproduced with permission from ref. 8.)

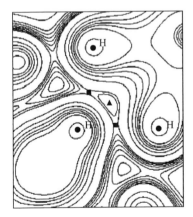

**Figure 9.7** Contour plot of the electron density for bifurcated S–H···(H–B)$_2$ dihydrogen bonds. The bond critical points are indicated as squares. (Reproduced with permission from ref. 8.)

In addition to this bifurcated dihydrogen bond, the SH group was involved in a classical C–H···S hydrogen bond, S(1)···H(2). Thus, this group acts simultaneously as a donor and an acceptor. When the SH group works as a hydrogen acceptor from the $C_c$–H bond through the sulfur atom, the S–H bond becomes more polar and as a consequence is a better donor than in the absence of a $C_c$–H···S interaction. Planas et al. believe that this *cooperative* effect, made possible by the SH group, plays an important role in the construction of two-dimensional supramolecular networks of the self-assembled neutral molecule *closo*-[3-Ru($C_6H_6$)-8-HS-1,2-$C_2B_9H_{10}$]. According to calculations, this hydrogen/bifurcated dihydrogen bonding demonstrates an energy gain of 1.4 kcal/mol. The same effects have been observed in the *closo*-[3-Co($C_5H_5$)-8-HS-1,2-$C_2B_9$ $H_{10}$] system.

The first experiments demonstrating the possibilities of dihydrogen bonds formed between ions $BH_4^-$ and the various NH proton donors as preorganizing interactions in the topochemical assembly of covalent materials due to solid-state $H_2$ elimination [eq. 9.1] were performed by Custelcean and Jackson in 1998 [9]. Further investigations in this field, directed to generalization of the strategy used in the controlled assembly of extended crystalline covalent networks, resulted in perspective molecular systems, one of which, **1**, is shown in Figure 9.8. The structure shows that solid complex **1** forms dimers that are held together by *four* classical hydrogen bonds complemented by *four* orthogonal dihydrogen bonds,

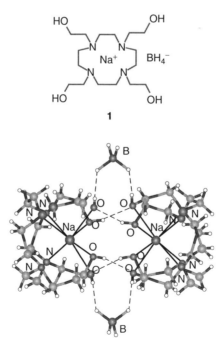

**Figure 9.8** Solid-state structure of NaBH$_4$ · THEC and self-assembly of this molecule into dihydrogen-bonded dimers. (Reproduced with permission from ref. 4.)

OH$\cdots$HB. One can assume that these dihydrogen bonds play a principal role in the formation of these dimer structures. Finally, the authors demonstrated that these dihydrogen bonds provide *preservation of crystallinity* during O–H$\cdots$H–B conversion into O–B bonds with $H_2$ elimination. In fact, solid-state decomposition of $NaBH_4 \cdot THEC$ and other systems leads to a crystalline covalent product by a crystal-to-crystal process.

## 9.1. CONCLUDING REMARKS

1. Similar to classical hydrogen bonding, dihydrogen bonds have a good combination of strength and directionality, potentially allowing rapid self-organization of molecular building blocks into extended regular structures.

2. In most cases, dihydrogen bonds do not show cooperative energy effects. However, even in the absence of cooperative effects, increasing the number of dihydrogen bonds leads to a large stabilization energy. This is a very important prerequisite for the creation of molecular assemblies.

3. Since hydrogen and dihydrogen bonds have comparable energies and directionality, their combination also holds interest for crystal engineers.

4. Dihydrogen bonds can be preorganizing interactions for the topochemical assembly of covalent materials due to solid-state $H_2$ elimination.

## REFERENCES

1. G. R. Desiraju, *J. Chem. Soc. Dalton Trans.* (2000), 3745.
2. J. G. Planas, F. Teixidor, C. Vinas, M. E. Light, M. B. Hursthouse, *Chem. Eur. J.* (2007), **13**, 2493.
3. J. P. Campbell, J. W. Hwang, V. G. Young, R. B. Dreele, C. J. Cramer, W. L. Gladfelter, *J. Am. Chem. Soc.* (1998), **120**, 521.
4. R. Custelcean, M. Vlassa, J. E. Jackson, *Chem. Eur. J.* (2002), **8**, 302.
5. S. Y. Liu, D. W. Michael, C. E. Dykstra, J. M. Lisy, *J. Chem. Phys.* (1986), **84**, 5032.
6. G. Merino, V. I. Bakhmutov, A. Vela, *J. Phys. Chem. A* (2002), **106**, 8491.
7. O. A. Filippov, A. M. Filin, N. V. Belkova, V. N. Tsupreva, Y. V. Schmirova, I. B. Sivaev, L. M. Epstein, E. S. Shubina, *J. Mol. Struct.* (2006), **790**, 114.
8. J. G. Planas, C. Vinas, F. Teixidor, A. Comas-Vives, G. Ujaque, A. Lledos, M. E. Light, M. B. Hursthouse, *J. Am. Chem. Soc.* (2005), **127**, 15976.
9. R. Custelcean, J. E. Jackson, *J. Am. Chem. Soc.* (1998), **120**, 12935.

# 10

# DIHYDROGEN BONDS AS INTERMEDIATES IN INTERMOLECULAR PROTON TRANSFER REACTIONS

Proton transfer to negatively charged hydrogen atoms has attracted the attention of many chemists over the last two decades. This process plays an important role in many chemical and biochemical phenomena that occur in the gas phase, in solution, and in the solid state [1–3]. For example, direct proton attack on hydride ligands generates transition metal dihydrogen complexes which are then involved in various catalytic transformations [4]:

$$M–H + HX \rightarrow [M(H_2)]^+X^- \qquad (10.1)$$

Stoichiometric and catalytic ionic hydrogenation as the sequence transfer of atoms $H^+$ and $H^-$ to olefins, ketones, or iminies,

$$H^+ + R_2C{=}O \rightleftharpoons R_2C^+–OH \qquad (10.2)$$
$$R_2C^+–OH + HM \rightarrow R_2CH–OH + M^+$$

has opened new methods of organic synthesis. Finally, proton transfer to $H^{\delta-}$ releases molecular hydrogen from materials used in hydrogen storage. It is obvious that mastering these processes is impossible without a detailed understanding of mechanisms in which proton transfer is a key step.

The role of classical hydrogen bonding in proton transfer to regular organic bases is well recognized. By analogy, dihydrogen bonds could be important organizing interactions or important intermediates, directing and assisting proton transfer to a negatively charged hydrogen atom and thus contributing to its kinetics. In earlier chapters we considered factors that affect the geometry and energy of dihydrogen bonds under conditions that imply the absence of full proton transfer. The latter can, however, lead to shorter lifetimes of H$\cdots$H bonds,

*Dihydrogen Bonds: Principles, Experiments, and Applications*, By Vladimir I. Bakhmutov
Copyright © 2008 John Wiley & Sons, Inc.

complicating their direct observation. However, the presence and transformations of such dihdyrogen-bonded states can be revealed kinetically. All these questions are in focus in this chapter, which begins with methods used in kinetic studies of proton transfers.

## 10.1. METHODS EMPLOYED IN KINETIC STUDIES OF PROTON TRANSFER TO HYDRIDIC HYDROGENS

Kinetics of proton transfers can be probed by any convenient physicochemical method having good resolution with respect to initial reagents and final products which allows us to trace quantitatively their mutual transformations. However, the choice of a specific technique depends on the nature of indicator groups or indicator properties in proton–hydride pairs and on the time scale inherent in the method chosen. Generally, the lifetimes of proton-accepting components in the presence of proton donors cover a region between $10^{-6}$ and $10^3$ s as a function of the acidic and basic strengths of the reagents used and, their concentrations, the nature of the solvents used, and temperature. It should be emphasized that protonation processes can be reversible or irreversible when, for example, full proton transfer leads to $H_2$ elimination. To avoid $H_2$ elimination and thus a distortion of the kinetic data, the reagents should be mixed at the lowest temperatures using a convenient procedure.

### 10.1.1. Electrochemical Experiments

Electrochemical measurements can be used as a kinetic tool when the oxidation potentials $E_i$ of the initial and final compounds are remarkably different. It has been found that transition metal dihydrogen complexes obtained by protonation of transition metal hydrides have more positive oxidation potentials. For example, the electrochemical potentials required for the oxidation of $FeH_2[P(CH_2CH_2PPh_2)_3]$ and $\{FeH(H_2)[P(CH_2CH_2PPh_2)_3]\}^+$ are measured as $-0.24$ and $0.9$ V, respectively. It is obvious that there is a wide range of potentials at which the initial hydride molecule is oxidized and the final dihydrogen complex remains stable [5]. This circumstance can be used to monitor the proton transfer process by measuring the time dependence of the limiting current at one of the intermediae values of the potential. However, in reality the situation is not so simple. For example, the appearance of ions $X^-$ in eq. (10.1) can complicate electrochemical experiments. In fact, $E_i$ values for the ions $Cl^-$, $Br^-$, and $CF_3CO_2^-$, which are often observed on reaction coordinates, are measured as $-0.24$, $-0.22$, and $-0.15$ V, respectively. They can obviously contribute to the limiting current at a potential that is similar to the $E_i$ value of a dihydrogen complex. For this reason, the proton transfer can be monitored better at a potential that is closer to the $E_i$ value of the initial dihydride.

Figure 10.1 illustrates a typical kinetic trace obtained by electrochemical experiments during protonation of the dihydride $FeH_2[P(CH_2CH_2PPh_2)_3]$. It is clear that such curves can then be treated in terms of an appropriate kinetic model.

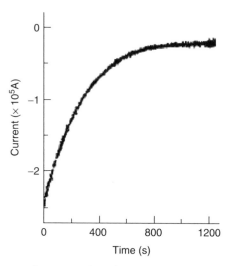

**Figure 10.1**  Kinetic trace for protonation of dihydride $FeH_2[P(CH_2CH_2PPh_2)_3]$ in THF with an excess of $HBF_4$ at room temperature. The intensity of current was measured at a platinum disk electrode maintained at $-0.1$ V with vigorous stirring. (Reproduced with permission from ref. 5.)

For example, since the experiments shown in Figure 10.1 have been performed under pseudo-first-order conditions, this curve can easily be treated by a single exponential [5].

### 10.1.2. Time-Dependent IR Spectra

The IR monitoring of proton transfer is particularly successful when the process is relatively slow and the molecules under investigation have good IR indicators. One such indicator is the CO group, which usually exhibits relatively narrow bands. Figure 10.2 illustrates the time-dependent IR experiments performed for a low-temperature protonation of hydride $CpRuH(CO)(PCy_3)$ with $(CF_3)_3COH$. As shown, the intensity of the CO band belonging to the initial hydride $CpRuH(CO)(PCy_3)$ decreases with time, providing an accurate kinetic analysis. Moreover, in contrast to many other spectral methods, the IR spectra show clearly the presence of various intermediates located on the reaction coordinate, the nature of which is discussed below. It should be emphasized again that the separate observation of such fast-transforming intermediates is a great advantage of IR spectroscopy. Finally, if the molecules investigated are active in visible–UV spectra, they can obviously be used in kinetic experiments.

### 10.1.3. NMR Spectra

Generally, $^1$H NMR spectra (or the spectra of other nuclei) show initial hydride molecules and products of their protonation with chemical shift differences that

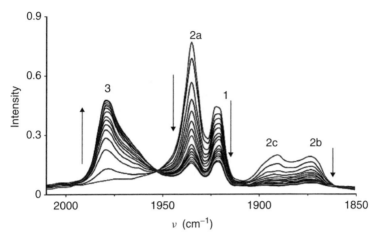

**Figure 10.2** Time-dependent IR spectra in the ν(CO) range of CpRuH(CO)(PCy$_3$) (compound **1**) (0.004 mol/L) in the presence of 6 equiv of (CF$_3$)$_3$COH at 220 K in hexane (over a 30-minute period). Compounds **2a** to **2c** are dihydrogen-bonded complexes, and **3** is a hydrogen-bonded ion pair. (Reproduced with permission from ref. 6.)

are generally sufficient to trace their mutual transformations. A typical example is the dihydride Re(PiPr$_3$)$_2$(NO)(CO)H$_2$, which exhibits two hydride resonances with the chemical shifts shown in Figure 10.3. A low-temperature mixing of this dihydride with a fourfold excess of CF$_3$COOH in toluene-d$_6$ leads to the appearance of a broadened resonance at −1.6 ppm and a narrow triplet resonance at −1.94 ppm, belonging to the dihydrogen and terminal ligands in the molecule

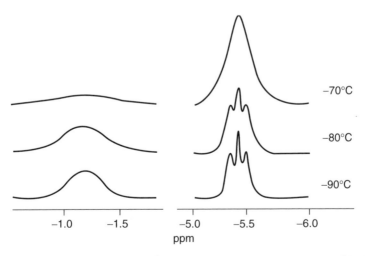

**Figure 10.3** Temperature-dependent $^1$H NMR spectra of in the presence of Re(PiPr$_3$)$_2$(NO)(CO)H$_2$ of CF$_3$COOH in toluene-d$_6$. (Reproduced with permission from ref. 7.)

$[Re(PiPr_3)_2(NO)(CO)H(H_2)]^+CF_3COO^-$ [7]. At smaller $CF_3COOH$ concentrations, both molecules are easily observed together. Under these conditions, by analogy with time-dependent IR experiments, such $^1H$ NMR spectra will be convenient for time monitoring of the protonation process if a proton transfer is slow on the NMR time scale (the usual NMR time scale refers to lifetimes between 1 and $10^{-6}$ s). $^{31}P\{^1H\}$ NMR spectra will also be appropriate in this context.

In addition, the NMR technique provides kinetic measurements for reversible proton transfers that are fast and occur on the NMR time scale (i.e., for processes with rates within $10^{-1}$ and $10^6$ s$^{-1}$). Under these conditions the NMR resonances undergo a *chemical exchange*. By definition [8], a chemical exchange between, for example, two nonequivalent spin states A and X,

$$
\begin{array}{ccc}
A & \rightleftharpoons & X \\
\nu^A; \tau^A & & \nu^X; \tau^X
\end{array}
\qquad (10.3)
$$

directly affects the line shapes of the A and X resonances when the exchange frequencies, $\nu_E$, are comparable with chemical shift differences $(\nu^A - \nu^X)$ measured in frequency units. Increasing the exchange rate leads to typical temperature evolution of resonances A and B: The exchanging lines broaden and coalescence. A fast exchange results in observation of a single resonance which is detected at an average frequency: $\nu = (\nu^A + \nu^X)/2$. Since the shape of lines in NMR is known analytically (this is the main advantage of NMR with respect to other spectroscopic methods), a full line-shape analysis, performed for the resonances observed, or approximate treatments of these lines in the slow- or fast-exchange limits, leads to the determination of the corresponding lifetimes, $\tau^A$ and $\tau^X$. These lifetimes can be transformed to the corresponding rate constants if the kinetic orders of the processes investigated are known. For example, the protonation of $Re(PiPr_3)_2(NO)(CO)H_2$ with $CF_3COOH$ at low temperatures does not lead to $H_2$ elimination and therefore is reversible. As shown in Figure 10.3, both hydride lines (and also the OH resonance of the acid, which is not shown in the figure) subsequently broaden with increasing temperature. Thus, protonation occurs on the NMR time scale, and line-shape analysis can be employed. Treatment of these spectra gives a proton transfer rate constant $k$ $(-80°C)$ of $4.4 \times 10^2$ L/mol·s and an activation energy $\Delta E^{\ddagger}$ of 11.5 kcal/mol, obtained in the framework of the bimolecular model.

## 10.1.4. Stopped-Flow Experiments

The stopped-flow technique, including tools for fast mixing of reagents, has an advantage when complete and even irreversible transformations of initial reagents to final products takes from 10 to $10^3$ s. Generally, fast-mixing units are coupled to a diode-array UV–visible spectrophotometer (or other spectroscopic instrument), including the software, which both supports the experiments and provides an analysis in the framework of the appropriate kinetic model. Figure 10.4 illustrates a typical room-temperature kinetic experiment carried out for protonation

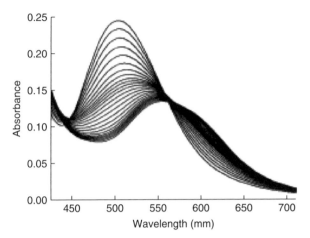

**Figure 10.4** Typical time-dependent spectral changes observed for the reaction of cluster [W$_3$S$_4$H$_3$(dmpe)$_3$]PF$_6$ with HCl in a CH$_2$Cl$_2$ solution at 25.0° C. The data were recorded for 1000 s with a logarithmic time base. (Reproduced with permission from ref. 9.)

of the cluster [W$_3$S$_4$H$_3$(dmpe)$_3$]PF$_6$ with HCl in CH$_2$Cl$_2$ solution [9]:

$$[W_3S_4H_3(dmpe)_3]^+ + 3HCl \rightarrow [W_3S_4Cl_3(dmpe)_3]^+ + 3H_2 \qquad (10.4)$$

It follows from the patterns that protonation is accompanied by clear spectral changes. These experiments have been analyzed in the framework of a kinetic model, where the first-order dependence of the rates observed for cluster concentrations were established by independent experiments.

Finally, it is worth mentioning that all the kinetic experiment described above should have good reproducibility.

## 10.1.5. Theoretical Approaches

Commonly, a chemical process can be represented as a series of several elementary reaction steps in which a transformation in every elementary step goes from a minimum on the potential energy surface, corresponding to the reagents $R_1$ and $R_2$, over an energy maximum corresponding to a transition state, to the next energy minimum, characterizing the products of the reaction, $P_1$ and $P_2$. This transformation is shown schematically in Figure 10.5. It is clear that the experimental kinetic measurements are directed to a determination of the rate and the height of the energy barrier. On the other hand, there is a definite possibility of determining the potential energy surface for the elementary reaction step by quantum mechanics. In terms of quantum mechanics, a stationary state of a molecular system is characterized by the solution of the many-electron Schrödinger equation,

$$\hat{H}_T \theta = E_T \theta \qquad (10.5)$$

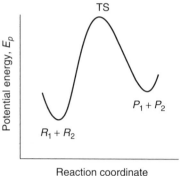

**Figure 10.5** Energy profile corresponding to the elementary reaction step transforming the initial reagents $R_1 + R_2$ to the products $P_1 + P_2$. (From ref. 10.)

where $\theta$ represents a wave function. In turn, this function is dependent on the electron and nuclear coordinates and $\hat{H}$ represents the Hamiltonian of the molecule with the energy of the system, $E$. If electrons move much faster than nuclei, eq. (10.5) is transformed to the reduced Schrödinger equation

$$\hat{H}\psi = E\psi \tag{10.6}$$

where $E$ is the potential energy of the molecule described by the nuclear coordinates only. Here $\hat{H}$ is the Hamiltonian of the molecule with the kinetic energy of the nuclei omitted, and $\psi$ is the electron wave function. The latter obviously depends on the coordinates of the electrons only.

The potential energy surface can now be obtained by the solution of eq. (10.6) with functions $\psi$ expanded in terms of the Slater determinants $D_i$:

$$\psi = \sum f_i D_i \quad \text{(from } i = 1 \text{ to } i = m_0) \tag{10.7}$$

These determinants are written

$$D_i = \varphi_{i1}(1)\varphi_{i2}(2)\cdots\varphi_{in}(n) \tag{10.8}$$

where the general elements for row $k$ and column $j$ are given by the function $\varphi_{ik}(j)$. The index $j$ shows the space and spin coordinates of electron $j$. In turn, the function $\varphi_{ik}(j)$ can be expressed as a linear combination of a set of functions $\chi_r(j)$,

$$\varphi_{ik}(j) = \sum C_{ikj}\chi_r(j) \tag{10.9}$$

known as *atomic orbitals*. In practice, the coefficients $f_i$ and $C_{ikj}$ are varied in eqs. (10.7) and (10.9) to reach a minimum of the energy $E$, which can be written

$$E = \frac{\int \psi^*\hat{H}\psi \, d\tau}{\int \psi^*\psi \, d\tau} \tag{10.10}$$

This is the first principle for determination of the potential energy surface.

Due to great progress in the theory and in computer technology during the past decade, the description of the potential energy surface for relatively small molecular systems can reach an accuracy close to chemical (around 1 kcal/mol) or an accuracy of ~5 kcal/mol for larger molecular systems. In addition, one of the more important advantages of computational chemistry is the ability to locate not only reagents or products of the reaction on a potential energy surface but also to characterize and investigate transition states.

It is well known that a solvent can cause dramatic changes in rates and even mechanisms of chemical reactions. Modern theoretical chemistry makes it possible to incorporate solvent effects into calculations of the potential energy surface in the framework of the continuum and explicit solvent models. In the former, a solvent is represented by a homogeneous medium with a bulk dielectric constant. The second model reflects specific molecule–solvent interactions. Finally, calculations of the potential energy surface in the presence or absence of solvents can be performed at various theory levels that have been considered in detail by Zieger and Autschbach [10].

## 10.2. PROTON TRANSFER TO A HYDRIDIC HYDROGEN IN THE SOLID STATE

According to theoretical studies, the dihydrogen-bonded complex $Li–H\cdots H–HF$ is transformed to the system $Li^+\cdots H–H\cdots F^-$, which is capable of $H_2$ elimination in the presence of an external electric field [11]. This result shows the possibility of proton-to-hydride transfer within crystals in the solid state. Due to the absence of intense molecular motions and additional intermolecular interactions, such a solid-state transformation of a dihydrogen-bonded system could be a good argument that proton transfer to hydridic hydrogen actually occurs via a dihydrogen bond. In accord with this statement, the long-term thermal decomposition of cyclotrigallazane, $[H_2GaNH_2]_3$, showing short $GaH\cdots HN$ distances, yields nanocrystalline gallium nitride [12].

Figure 10.6 illustrates the x-ray structure of $N$-[2-(6-aminopyridyl)]acetamidine (NAPA) cyanoborohydride synthesized and reported by Custelcean and Jackson [13]. According to the x-ray data, this compound exists as two independent centrosymmetric dimers in the unit cell. As shown, the two dimers exhibit short $H\cdots H$ contacts: 1.98, 2.12, and 2.26 Å and 2.04, 2.09, and 2.31 Å, respectively. All the contacts are smaller than the van der Waals sum of 2.4 Å. The $NH\cdots H–B$ angles are remarkably bent, changing between 91.6 and 126.3°. The $N–H\cdots HB$ angles are larger and vary between 148.1 and 175.6°. Thus, the geometry and

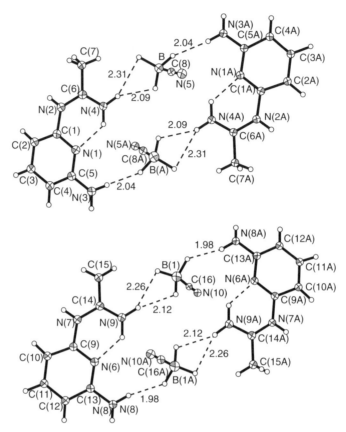

**Figure 10.6**  X-ray crystal structure of N-[2-(6-aminopyridyl)]acetamidine cyanoboro-hydride. The dashed lines show the short H··· H contacts. (Reproduced with permission from ref. 13.)

the H··· H distances of the system correspond closely to dihydrogen bonding. Custelcean and Jackson have found that the solid-state decomposition of this dihydrogen-bonded system leads to the compound shown in Structure 10.1, which can be considered a product of proton transfer. Scheme 10.1 is a more common scheme for such transfers, where $H_2$ elimination leads to the formation of new covalent chemical bonds.

Later, this mechanism was described in greater detail by the kinetic study of solid-state transformation from the dihydrogen-bonded system $LiBH_4 \cdot$ triethanolamine to a covalent bonded material [14]. Variable-temperature experiments performed by [11]B solid-state magic-angle spinning NMR, Fourier Transform IR, X-ray diffraction, and optical microscopy have resulted in rate constants of $37 \times 10^{-4}$ to $111 \times 10^{-4}$ min$^{-1}$ in the temperature range 105 to 120° C. These constants, via their temperature dependence (Figure 10.7), give an activation energy barrier of $21.0 \pm 2.4$ kcal/mol. Treatment of the constants by

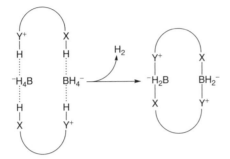

**Structure 10.1**

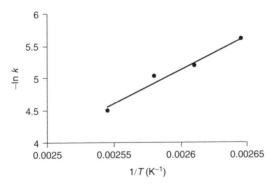

**Scheme 10.1**

**Figure 10.7** Arrhenius plot for solid-state proton transfer in dihydrogen-bonded complex LiBH$_4$ · triethanolamine. The data are obtained in the framework of the Avrami–Erofeev model. (Reproduced with permission from ref. 14.)

the well-known equation

$$k = \frac{kT}{h} \exp\left(\frac{-\Delta H^{\ddagger}}{RT}\right) \exp\left(\frac{\Delta S^{\ddagger}}{R}\right) \tag{10.11}$$

leads to $\Delta H^{\ddagger}$ and $\Delta S^{\ddagger}$ values of $20.1 \pm 2.4$ kcal/mol and $-16.8 \pm 6.2$ eu, respectively.

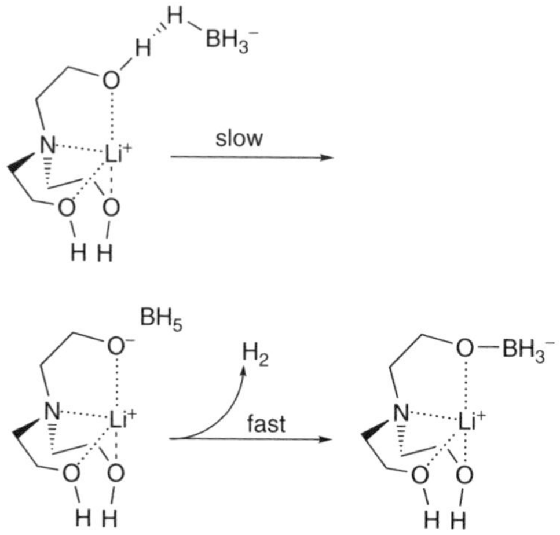

**Scheme 10.2**

Very important experiments with the labeled system $LiBH_4 \cdot N(CH_2CH_2OD)_3$ have shown that the proton transfer is slow and a rate-determining step of the transformation, as demonstrated in Scheme 10.2. Finally, it should be noted that the activation energy parameters found for solid-state proton transfer are comparable with those measured in solution.


## 10.3. PROTON TRANSFER TO A HYDRIDE LIGAND IN SOLUTIONS OF TRANSITION METAL HYDRIDE COMPLEXES: THEORY AND EXPERIMENT

The focuses of this section are transition metal hydride systems that transform to dihydrogen complexes in solution. In this context, only an intramolecular version of proton transfer to a hydrogen ligand is relatively simple and phenomenologically similar to the solid-state process considered above. For example, when a $CD_2Cl_2$ solution of the Ir · hydride complex with well-established intramolecular dihydrogen bonds (see Structures 5.9) is exposed for 5 minutes under the gas $D_2$, the intensities of the NH and IrH resonances in the $^1H$ NMR spectra decrease strongly, due to an H/D exchange. None is observed in THF, destroying these bonds [15]. These results are easy explained by proton transfer along the dihydrogen bond with the formation of a dihydrogen complex that exchanges rapidly with $D_2$ gas (Scheme 10.3).

On the other hand, it is well known that a dihydrogen complex can be generated through direct proton attack on a hydride ligand or initial protonation of a metal center, leading to a new classical dihydride, $[MH_2]^+$, which then converts to the dihydrogen complex $[M(\eta^2-H_2)]^+$. The latter is a thermodynamic product of the protonation reaction shown in Scheme 10.4 [16,17].

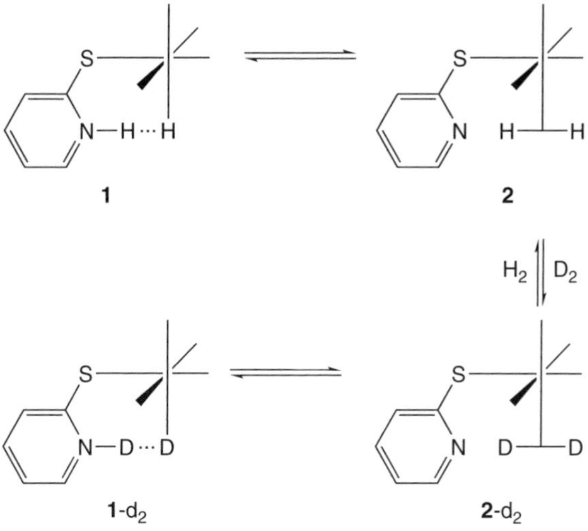

**Scheme 10.3**

$$M-H + HX \longrightarrow [M(\eta^2-H)_2]^+ [X]^- \longrightarrow M-X + H_2 \qquad (1)$$

$$M-H + HX \longrightarrow [MH_2]^+ [X]^- \longrightarrow [M(\eta^2-H)_2]^+ [X]^- \qquad (2)$$

$$M-X + H_2 \longleftarrow$$

**Scheme 10.4**

It should be emphasized that these two schemes are principally different. The second mechanism corresponds to a situation in which the formation of $M \cdots H$ bonds precedes the full proton transfer, while dihydrogen bonds can be formed only via direct protonation of the hydridic hydrogen. In addition, the protonation of the metal center can be slow at very fast transformation to the dihydrogen complex, masking detection of the dihydride molecule. Therefore, first, it is necessary to show what hydride center is protonated faster.

Taking into account the importance of this problem, let us follow the evidence suggested by Papich and co-workers [18]. In the presence of various acids, the authors investigated the $^1H$ NMR behavior of the hydride $CpW(CO)_2(PMe_3)H$, which exists in solution as a mixture of two isomers, **3c** and **3t**, depicted in Scheme 10.5. If the hydride ligand in complex **3** is actually protonated faster, the hydride–proton exchange with an acid will be faster than protonation of the W center, to give the cationic dihydride **4**. This process is illustrated in Scheme 10.6. Qualitative evidence for this statement follows directly from the $^1H$ NMR spectrum in Figure 10.8. As shown, the hydride resonance of **3t** is *selectively broadened* due to the H/H exchange with the acid on the $^1H$ NMR time scale.

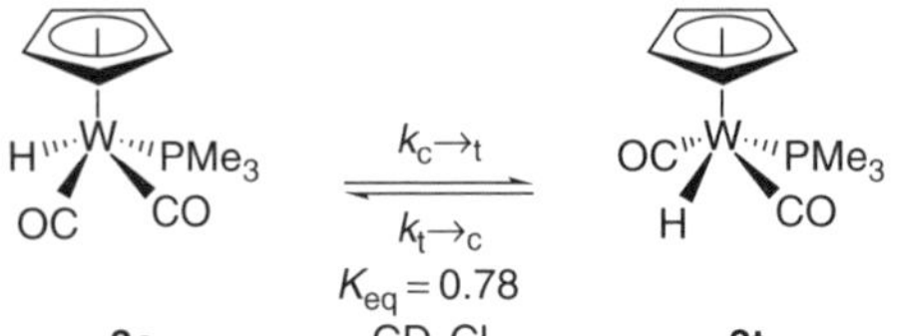

**Scheme 10.5**

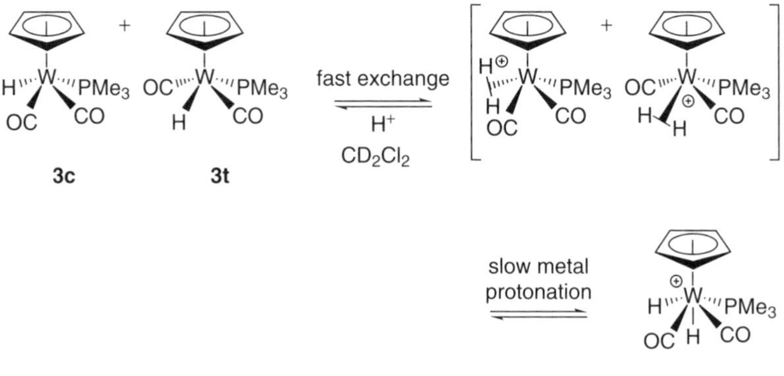

**Scheme 10.6**

This effect is the result of selective protonation of the hydride ligand of **3t** because, according to Papich et al., "there is no plausible reason for metal protonation to be faster for **3t** than **3c**."

Quantitative treatment of rate constants for the hydride attack, $k_{HP}$, the metal protonation, $k_{FPM}$, and the exchange process in the framework of Scheme 10.7 have resulted in $k_{HP} = 2.7 \times 10^{-3}$ M$^{-1}$/s and $k_{FPM} = 2.8 \times 10^{-4}$ M$^{-1}$/s. Thus, the hydride protonation occurs faster by a factor of 10. In earlier chapters we have shown that transition metal hydrides form dihydrogen bonds in the presence of proton donors. Now, based on the principle of microscopic reversibility, one can suggest that proton transfer to a hydridic hydrogen actually occurs via a dihydrogen bond.

### 10.3.1. Kinetic Experiments with Proton Transfer to a Hydride Ligand in Solution

As we have shown in Chapter 2, strong classical hydrogen bonds dominate in the protonation of the common organic bases. For example, proton transfer from acetic acid, H–A, to pyridine, B, occurs via the various hydrogen-bonded complexes in Structures 10.2, which can be observed directly by $^1$H and $^{15}$N NMR spectroscopy in CDClF$_2$/CDF$_3$ at low temperatures [23]. In addition, it is well known that the hydrogen bond strength increases with partial proton transfer.

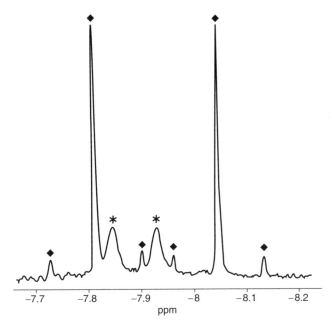

**Figure 10.8** $^1$H NMR of the hydride region of complex **3** (0.019 M) (Scheme 10.5) broadened by exchange with [PhNH$_3$·(OEt$_2$)$_{1-2}$][B(Ar$^f$)$_4$] (0.0084 M) in CD$_2$Cl$_2$ at 263 K. Signals marked with asterisks and diamonds are due to **3t** and **3c**, respectively. (Reproduced with permission from ref. 18.)

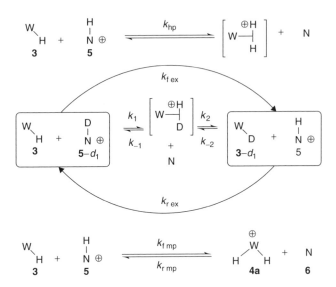

**Scheme 10.7** Kinetic scheme showing the relationships between the protonation and H/D exchange process for the W hydride in the presence of an NH acid. (Reproduced with permission from ref. 18.)

$$B \cdots H \cdots A$$

$$B^{\delta +} \cdots H \cdots (A \cdots H \cdots A)^{\delta -}$$

$$B^{\delta +} \cdots H \cdots (A \cdots H \cdots A \cdots H \cdots A)^{\delta -}$$

**Structure 10.2**

Then, if the proton overcomes an energy barrier due to its motion only, the barrier must be very small. In full agreement with this statement, protonation of regular organic bases is a very fast process.

Protonation of heteroatom Y in an organic base consists of a proton shift along a linear hydrogen bond, $O–H \cdots Y$, with the formation of an intimate ion pair, $O^- \cdots H–Y^+$, as the product of partial proton transfer. Then this hydrogen-bonded ion pair converts to a solvent-separated ion pair to give free ions if the solvent polarity is sufficient for full proton transfer. Since classical hydrogen bonds and dihydrogen-bonded complexes are similar, a similar concept can be used for proton transfer to hydride ligands. Here, as shown in Schemes 10.8, the reaction starts with the formation of dihydrogen bonds at the ends with the formation of solvent-separated ion pairs due to the relatively small permittivities of the solvents [tetrahydrofuran (THF), toluene, $CH_2Cl_2$] usually employed in the protonation of transition metal hydrides.

Table 10.1 lists kinetic data for the protonation of transition metal hydrides obtained through various spectroscopic (NMR, UV, IR) and electrochemical methods using stopped-flow mixing techniques. Most kinetic experiments have

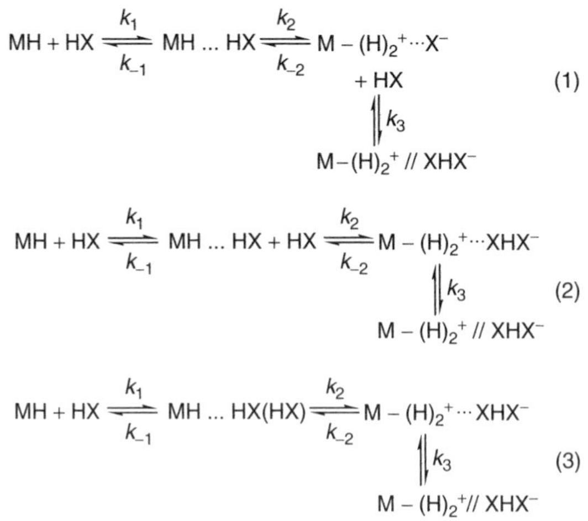

$$MH + HX \underset{k_{-1}}{\overset{k_1}{\rightleftharpoons}} MH \ldots HX \underset{k_{-2}}{\overset{k_2}{\rightleftharpoons}} M – (H)_2^+ \cdots X^-$$
$$+ HX$$
$$\Big\Updownarrow k_3 \qquad (1)$$
$$M – (H)_2^+ \,/\!/\, XHX^-$$

$$MH + HX \underset{k_{-1}}{\overset{k_1}{\rightleftharpoons}} MH \ldots HX + HX \underset{k_{-2}}{\overset{k_2}{\rightleftharpoons}} M – (H)_2^+ \cdots XHX^-$$
$$\Big\Updownarrow k_3 \qquad (2)$$
$$M – (H)_2^+ \,/\!/\, XHX^-$$

$$MH + HX \underset{k_{-1}}{\overset{k_1}{\rightleftharpoons}} MH \ldots HX(HX) \underset{k_{-2}}{\overset{k_2}{\rightleftharpoons}} M – (H)_2^+ \cdots XHX^-$$
$$\Big\Updownarrow k_3 \qquad (3)$$
$$M – (H)_2^+ \,/\!/\, XHX^-$$

**Scheme 10.8**  Proton transfer to a hydride ligand via dihydrogen bonds with participation of two proton-donor HX molecules: the dimer $(HX)_2$ forms the dihydrogen bond [pathway (3)]; the second HX molecule initiates the formation of the solvent-separated or contact ion pair [pathway (1) or (2), respectively].

**TABLE 10.1. Second-Order Rate Constants ($k$, 25° C) for Protonation of Transition Metal Hydrides with Formation of Dihydrogen Complexes in Solution**

| Complex | HX | $k$(L/mol·s) | Solvent | Ref. |
|---|---|---|---|---|
| [FeH$_2$PP$_3$] | HBF$_4$ | $1.7 \times 10^{-4}$ | THF | 5 |
| | CF$_3$COOH | $0.112 \times 10^2$ | | |
| | CF$_3$SO$_3$H | $0.176 \times 10^2$ | | |
| | HCl | $1.32 \times 10^{-2}$ | | |
| | HBr | $3.4 \times 10^{-2}$ | | |
| [FeH$_2$(dppe)$_2$] | HBF$_4$ | $9.7 \times 10^{-3}$ | THF | 19 |
| | CF$_3$COOH | $1.39 \times 10^{-2}$ | | |
| | CF$_3$SO$_3$H | $2.1 \times 10^{-2}$ | | |
| | HCl | $4.8 \times 10^{-2}$ | | |
| [CpW(CO)$_2$(PMe$_3$)H] | [Me$_2$NHC$_6$H$_4$ C(Me$_3$)]$^+$[BF$_4$]$^-$ | $2.7 \times 10^{-3a}$ | CD$_2$Cl$_2$ | 18 |
| [CpRuH(dppm)] | HBF$_4$ | $1.86 \times 10^2$ | THF | 20 |
| [CpRuH(dppe)] | HBF$_4$ | $0.70 \times 10^2$ | THF | 20 |
| [CpRuH(PPh$_3$)$_2$] | HBF$_4$ | $1.69 \times 10^2$ | THF | 20 |
| [RuH$_2$(dppe)$_2$] | HBF$_4$ | $1.1 \times 10^3$ | THF | 21 |
| | CF$_3$COOH | $9.2 \times 10^4$ | | |
| | HCl | $1.7 \times 10^6$ | | |
| [ReH$_2$(NO)(CO)($^i$Pr$_3$)$_2$] | CF$_3$COOH | $4.4 \times 10^{2b}$ | CD$_2$Cl$_2$ | 7 |
| [ReH$_2$(NO)(CO)(PMe$_3$)$_2$] | CF$_3$COOH | $192 \times 10^{2c}$ | CD$_2$Cl$_2$ | 7 |
| [Cp*Fe(dppe)H] | CF$_3$COOH | very fast | CH$_2$Cl$_2$ | 22 |
| | PFTB | $1.56 \times 10^2$ | CH$_2$Cl$_2$ | |
| | HFIP | 5.4 | CH$_2$Cl$_2$ | |
| | CF$_3$CH$_2$OH | $1.5 \times 10^{-3}$ | CH$_2$Cl$_2$ | |

[a] At −10° C.
[b] At −80° C.
[c] At −90° C.

been performed under pseudo-first-order conditions when acids have been added in large excess [5,19–22]. By contrast, [ReH$_2$(NO)(CO)(PR$_3$)$_2$]/CF$_3$COOH systems have been studied with comparable concentrations of reagents [7]. In all cases the overall kinetic order has been found to be close to 2.

The data in Table 10.1 illustrate the dependencies of rate constants on the nature of the acid used. For example, the rate of proton transfer to [Cp*Fe(dppe)H] increases strongly with the acidity of the proton donor as CF$_3$CH$_2$OH < (CF$_3$)$_2$CHOH < (CF$_3$)$_3$COH < CF$_3$COOH [22]. At the same time, the cis-hydride [FeH$_2$PP$_3$] shows very insignificant changes in the $k$ values in the order CF$_3$COOH < HCl < HBr [5]. The rate constants are smaller for the stronger acids HBF$_4$ and CF$_3$SO$_3$H. The same effect has been observed for [RuH$_2$(dppe)$_2$] [21] and [ReH$_2$(NO)(CO)(PMe$_3$)$_2$] protonated with CF$_3$COOH and [{3,5-(CF$_3$)$_2$-C$_6$H$_3$}$_4$B]$^-$[H(OEt$_2$)]$^+$ [7]. Thus, one can think that in common cases, the rate of proton transfer and the acid strength do not correlate. This result is easy explained by the relatively small permittivities of the solvents (THF, toluene, $CH_2Cl_2$). Even strong acids exist in an equilibrium between molecular forms, ion pairs, and free ions:

$$H-X \rightleftharpoons H^+//X^- \rightleftharpoons H^+ + X^- \tag{10.12}$$

In fact, if free $H^+$ attacks a hydride, proton transfer should be suppressed by the addition of $X^-$ ions. However, $NBu_4BF_4$, for example, does not affect the protonation rates of cis-$[FeH_2PP_3]$ by the action of $HBF_4$ in THF solutions. This is good evidence for the parallel attack of the molecular form HX and the ion pair under conditions when ion pairs react more slowly than HX [5].

Kinetic experiments with the acids HX and DX, performed under pseudo-first-order conditions, have revealed inverse kinetic isotope effects, $k_H/k_D$, shown in Table 10.2. These results can be interpreted either as the kinetic effects measured for a single-step proton transfer or as the inverse thermodynamic isotope effects in fast preequilibria [5]. Unfortunately, in contrast to the isotope effects measured for classical hydrogen bonds [24], the effects of deuterium on the thermodynamics of dihydrogen bonding are still unknown.

On the other hand, it is obvious that under pseudo-first-order conditions, the position of the kinetic preequilibrium should be shifted completely toward dihydrogen bonds. Therefore, the hypothesis of a single-step proton transfer via a late transition state could be successful for interpreting the isotopic effects observed. A structure similar to that of the contact ion pairs (see below) with almost complete formation of new bonds could represent such a transition state. The $k_H/k_D$ values can then be calculated by [5]

$$\frac{k(H)}{k(D)} = \exp\left\{7.06 \times 10^{-4}\left[\sum_i (\nu_i(H)) - \sum_i (\nu_i^{\ddagger}(H))\right]\right\} \tag{10.13}$$

**TABLE 10.2. Kinetic Isotope Effects for Protonation of Transition Metal Hydrides in Solution**

| Complex | HX | Solvent | $k(H)/k(D)$ | | Ref. |
| | | | Measured | Calculated | |
|---|---|---|---|---|---|
| $[FeH_2(dppe)_2]$ | $CF_3SO_3H$ | THF | 0.21 (25° C) | 0.87 | [19] |
| | HCl | | 0.36 (25° C) | 0.47 | |
| | HBr | | 0.55 (25° C) | 0.39 | |
| $[FeH_2PP_3]^a$ | $CF_3SO_3H$ | THF | 0.45 (25° C) | 0.87 | [5] |
| | HCl | | 0.62 (25° C) | 0.47 | |
| | HBr | | 0.64 (25° C) | 0.39 | |
| $[RuH_2(dppe)_2]$ | $CF_3COOH$ | THF | 0.80 (25° C) | 0.87 | [21] |
| | HCl | | | 0.38 (25° C) | 0.47 | |
| $[ReH_2(NO)(CO)(P^iPr_3)_2]$ | $CF_3COOH$ | $CD_2Cl_2$ | 1.4 (−80)°C | 0.87 | [7] |

$^a$PP$_3$ = $P(CH_2CH_2PPh_2)_3$.

where $v_i(H)$ and $v_i^{\ddagger}(H)$ are the stretching frequencies in the fundamental and transition states, respectively. Table 10.2 illustrates good agreement between the theoretical and experimental $k_H/k_D$ values.

As in the case of classical organic bases, proton transfer to a hydride ligand is not a single-step process and occurs via dihydrogen-bonded intermediates, where dihydrogen bonding is a weak preorganizing interaction. The latter cannot be responsible for the very large variations in rate constants of proton transfers shown in Table 10.1. For example, the $k$ constant measured for proton transfer from $CF_3COOH$ to $[FeH_2PP_3]$, $[ReH_2(NO)(CO)(PMe_3)_2]$, and $[RuH_2(dppe)_2]$ increases by *six orders*: from $0.112 \times 10^{-2}$ (25°C) to $192 \times 10^2$ ($-90°C$) to $9.2 \times 10^4$ (25° C) L/mol. In other words, these data show clearly that the formation of dihydrogen bonds cannot be a *rate-determining* step.

Scheme 10.8 represents proton transfer to hydride ligands with the participation of two proton donor molecules, emphasizing the role of homoconjugated $[X\cdots H\cdots X]^-$ species in the kinetics of the process. Note that the second HX molecule initiates the formation of the solvent-separated or contact ion pair, corresponding to pathway (1) or (2). Following the principles of formal kinetics, pathway (1) can be expressed via

$$d\frac{[MH]}{dt} = -\frac{k_1 k_2 k_3 [MH][HX]^2}{k_{-1}k_{-2} + k_{-1}k_3[HX] + k_2 k_3[HX]} \tag{10.14}$$

Since the formation of contact ion pairs seems to be rate-determining (i.e., $k_2 \ll k_{-1}$ and $k_3[HX] \gg k_{-2}$), eq. (10.14) is transformed to

$$\frac{d[MH]}{dt} = -\frac{k_1}{k_{-1}}k_2[MH][HX] \tag{10.15}$$

and thus the reaction represents a *first-order* process with respect to the acid. It is easy to show that under pseudo-first-order conditions (when the preequilibrium is shifted completely to $H\cdots H$ complexes), the measured rates become independent of HX concentrations.

Experimentally, this pathway has been well established from IR spectra of the $[CpRuH(CO)(PCy_3)]/(CF_3)_3OH$ system in $CH_2Cl_2$, where large variations in hydride/alcohol ratios did not affect slow transformation of the $H\cdots H$ complexes to hydrogen-bonded ion pairs with $k$ values between $1.4 \times 10^{-3}$ and $1.6 \times 10^{-3}$ s$^{-1}$ [25]. Activation parameters for this step (Table 10.3) have been determined in hexane [6]. It is probable that a similar mechanism operates for protonation of the hydrides $[ReH_2(NO)(CO)(PR_3)_2]$ with $CF_3COOH$ (Table 10.3) in $CD_2Cl_2$, where the reaction corresponds to first-order kinetics on the acid at hydride/acid ratios $\geq 1$ [7].

In contrast to the data above, UV monitoring of the hydride $[Cp^*Fe(dppe)-H]$ protonated with a 30- to 200-fold excess of $(CF_3)_3COH$ or $(CF_3)_2CHOH$ ($CH_2Cl_2$, pseudo-first-order conditions) led to the *first-order* dependence on alcohol concentration of the rates measured [22]. In the framework of pathway (1), the result

**TABLE 10.3. Activation Parameters for Proton Transfer to Hydride Ligands in Solution**

| System | $\Delta H^{\ddagger}$ (kcal/mol) | $\Delta S^{\ddagger}$ (eu) | Solvent | Ref. |
|---|---|---|---|---|
| CpRuH(CO)(PCy$_3$)/PFTB (1:2) | 11.0 | −19 | Hexane | 6 |
| ReH$_2$(NO)(CO)(P$^i$Pr$_3$)$_2$/CF$_3$COOH (1:1) | 11.1 | 12.0 | CD$_2$Cl$_2$ | 7 |
| ReH$_2$(NO)(CO)(PMe$_3$)$_2$/CF$_3$COOH (1:1) | 13.5 | 36.8 | CD$_2$Cl$_2$ | 7 |
| ReD(PMe$_3$)$_4$(CO)/CF$_3$COOD (1:1) | 14.4 | 14.8 | CH$_2$Cl$_2$ | 7 |

can be interpreted according to

$$\frac{d[\text{MH}]}{dt} = -\frac{k_1}{k_{-1}}\frac{k_2}{k_{-2}}k_3[\text{MH}][\text{HX}]^2 \tag{10.16}$$

obtained from eq. (10.14) by assuming that $k_3[\text{HX}] \ll k_{-2}$. It is clear that under pseudo-first-order kinetic measurement conditions, this equation will show a first-order dependence on the acid concentration.

On the other hand, the data above can be rationalized in the context of pathway (2), where a slow transformation of the H$\cdots$H complexes to contact ion pairs is accompanied by the second HX molecule in

$$\frac{d[\text{MH}]}{dt} = -\frac{k_1 k_2 k_3[\text{MH}][\text{HX}]^2}{k_{-1}k_{-2} + k_{-1}k_3 + k_2 k_3[\text{HX}]} \tag{10.17}$$

Then, at $k_2\,[\text{HX}] \ll k_{-1}$, eq. (10.17) is converted to

$$\frac{d[\text{MH}]}{dt} = -\frac{k_1 k_2 k_3[\text{MH}][\text{HX}]^2}{k_{-1}(k_{-2} + k_3)} \tag{10.18}$$

The latter will correspond to the first-order process under pseudo-first-order kinetic measurement conditions. As we show below, proton transfer in the [CpRu(CO)(PH$_3$)H]/CF$_3$COOH system is energetically preferable when the second HX molecule participates in each step of the process [6]. If the acid is initially a dimer, this situation corresponds, in kinetic terms, to

$$\frac{d[\text{MH}]}{dt} = -\frac{k_1 k_2 k_3[\text{MH}][\text{DM}]}{k_{-1}k_{-2} + k_{-1}k_3 + k_2 k_3} \tag{10.19}$$

where $(\text{HX})_2 = \text{DM}$, showing first-order kinetics for the DM acid. Due to the dimerization of CF$_3$COOH [26], this scheme cannot be excluded in interpreting the kinetic data collected for the [ReH$_2$(NO)(CO)(PR$_3$)$_2$]/CF$_3$COOH (1 : 1) system in CD$_2$Cl$_2$ [7]. It is clear that only detailed kinetic measurements for large variations in reagent concentrations can distinguish pathways 1 to 3.

A very interesting and complex protonation mechanism has been suggested for the hydride cluster $[W_3S_4H_3(dmpe)_3]PF_6$ in $CH_2Cl_2$ solutions. In the presence of an excess of HCl, a careful kinetic study of the process in eq. (10.4) by the stopped-flow technique [9] has revealed *three kinetically distinguishable steps*: very fast, fast, and slow, with rate constants $k_1$, $k_2$, and $k_3$. The kinetic order in the initial hydride cluster in the slow step has been measured as 1. At the same time, rate constants $k_1$ and $k_2$ have corresponded to a second-order dependence on acid concentration, while the third step has shown a zero kinetic order on HCl. The rate constants have been determined as $k_1 = 2.41 \times 10^5$ $M^{-2}$/s, $k_2 = 1.03 \times 10^4$ $M^{-2}$/s, $k_3 = 4 \times 10^{-3}$ $s^{-1}$. Note that the protonation process becomes simple at lower concentrations of HCl. Under these conditions it shows a single step with a first kinetic order on the acid.

The data have been interpreted in terms of two competitive mechanisms. One of them corresponds to a direct proton transfer along the dihydrogen bond $W-H\cdots H-Cl$ with the slow formation of $[W_3S_4Cl_3(dmpe)_3]^+$ and $H_2$. The pathway at high acid concentrations requires a second HCl molecule, forming a dihydrogen-bonded system:

$$W-H + 2HCl \rightleftharpoons W-H\cdots H-Cl\cdots H-Cl \qquad (10.20)$$

Then this adduct yields $[W_3S_4Cl_3(dmpe)_3]^+$, HCl, and $H_2$ in the rate-determining step. It should be emphasized that this mechanism and pathway (2), in Scheme 10.8 are different. In fact, the latter requires HX assistance to stabilize the ionic pair $M-(H_2)^+\cdots(XHX)^-$. According to the kinetic data presented in this section, such mechanisms are rather rare; pathway (1) in Figure 10.8 can be taken as a general mechanism of proton transfer.

## 10.3.2. Proton–Hydride Exchange as a Measure of Proton Transfer to a Hydride Ligand in Solution

Proton–hydride exchanges similar to those in Scheme 10.9 are manifested in the $^1H$ NMR spectra as broadenings and coalescences of the hydride and HX resonances upon heating and can be characterized quantitatively with the help of line-shape analysis or saturation transfer experiments. As in the case of proton

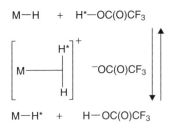

**Scheme 10.9**

**TABLE 10.4. Kinetic Parameters of Proton–Hydride Exchanges Determined by $^1$H NMR in Solution**

| System | $k$(exch)(L/mol·s) [at $T°C$] | $\Delta H^{\ddagger}$ (kcal/mol) | $\Delta S^{\ddagger}$(eu) | Solvent |
|---|---|---|---|---|
| [CpW(CO)$_2$(PMe$_3$)H]/ | | | | |
|   [(CH$_3$)$_3$C$_6$H$_4$N(CH$_3$)$_2$D]$^+$ | $1.8 \times 10^{-4}(-10)$ | 10.7 | $-35.2$ | CD$_2$Cl$_2$ |
| [CpW(CO)$_2$(PMe$_3$)D]/ | | | | |
|   [(CH$_3$)$_3$C$_6$H$_4$N(CH$_3$)$_2$H]$^+$ | $8.9 \times 10^{-4}(-10)$ | 11.2 | $-30.2$ | CD$_2$Cl$_2$ |
| [ReH$_2$(NO)(CO)(P$^i$Pr$_3$)$_2$]/ | | | | |
|   CF$_3$COOH | $33.7 \times 10^2 \ (-70)^a$ | 7.7 | $-2.3$ | Toluene-d$_8$ |
| [ReD$_2$(NO)(CO)(P$^i$Pr$_3$)$_3$]/ | | | | |
|   CF$_3$COOD | $16.0 \times 10^2 \ (-70)^a$ | — | — | Toluene |
| [ReH$_2$(NO)(CO)(PO$^i$Pr$_3$)$_2$]/ | | | | |
|   CF$_3$COOH | $20.7 \times 10^2 \ (-66)^{a,b}$ | 7.4 | $-5.5$ | CD$_2$Cl$_2$ |
| [ReD$_2$(NO)(CO)(PO$^i$Pr$_3$)$_2$]/ | | | | |
|   CF$_3$COOD | $4.6 \times 10^2 \ (-70)^a$ | 7.1 | $-10.3$ | Toluene |
| [ReD$_2$(NO)(CO)(PO$^i$Pr$_3$)$_2$]/ | | | | CD$_2$Cl$_2$ |
|   CF$_3$COOD | $3.6 \times 10^2 \ (-70)^a$ | 6.0 | $-16.6$ | /toluene(1 : 1) |
| | | | | CD$_2$Cl$_2$ |
| | $0.93 \times 10^2 \ (-60)^b$ | 7.7 | $-13.9$ | /toluene(1 : 1) |
| [ReH$_2$(NO)(CO)(PMe$_3$)$_2$]/ | | | | |
|   CF$_3$COOH | $250 \times 10^2 \ (-73)^a$ | 7.8 | 0 | Toluene-d$_8$ |
| [ReD(PMe$_3$)$_4$(CO)]/ | | | | |
|   CF$_3$COOD(1 : 1) | $0.92 \times 10^2 \ (-60)$ | 6.3 | $-19.2$ | CH$_2$Cl$_2$ |

*Source:* Data from refs. 7 and 18.

$^a$Trans to the NO group.

$^b$Cis to the NO group.

transfer, the exchange rates in Table 10.4 depend strongly on the nature of the hydrides and are sensitive to isotope displacement. The overall kinetic order of the exchanges is also equal to 2 [7,18], in agreement with almost-negative activation entropies. The nature of the acids also has a significant influence on exchange rates: The exchange is faster when the anion of HX is capable of assistance [7].

As we have mentioned, the proton–hydride exchange is regioselective in an isomeric mixture of the hydride [CpW(CO)$_2$(PMe$_3$)–H] in the presence of [PhNH$_3$(OEt$_2$)]$^+$$_{1-2}$[B(Ar$^f$)$_4$]$^-$, the proton–hydride exchange is fast only for the *trans* isomer [18]. The exchanges are highly regioselective in [d$_8$]-toluene solutions of the [ReH$_2$(NO)(CO)(PR$_3$)$_2$]/CF$_3$COOH system, where only the H ligands located trans to NO are involved in the exchange (Table 10.4) [7]. Since this effect correlates with the regioselective dihydrogen bonding and bonds

Re$\cdots$HOR are not observed, the dihydrogen-bonded adducts can be taken as a starting point for the proton–hydride exchange process. A priori, a hydride atom and an acidic proton can exchange their positions if both hydrogen atoms are bound to a metal center in a transition state or an intermediate. It is clear that the solvent-separated ion pairs, $M(H_2)^+//X^-$, could play the roles of such states. Experimentally, this statement can be supported by the equal rates measured for the proton–hydride exchange and the formation of dihydrogen complexes. This case has been well documented for the $[ReH_2(NO)(CO)(P^iPr_3)_2]/CF_3COOH$ system in the relatively polar $CD_2Cl_2$: [7] line-shape analysis of the hydride resonances (a test for the exchange) and $^{31}P$ NMR signals of $[ReH_2(NO)(CO)(P^iPr_3)_2]$ and $[ReH(H_2)(NO)(CO)(P^iPr_3)_2]^+[CF_3COO]^-$ (a test for the formation of dihydrogen complexes) gave very similar rate constants (15 to $20 \times 10^2$ L/mol·s at $-70°C$). It is clear that under these conditions, the proton–hydride exchange is a good probe for full proton transfer to a hydride ligand. However, as in the case of the usual organic bases, this situation can change. In fact, proton exchanges in organic acid–base pairs can occur without charge separation via the cyclic nonpolar transition states in Structure 10.3, where N–H is a base and H–O is an acid.

As a result, for example, the dihydrogen complexes $[ReH(H_2)(CO)(NO)(PR_3)_2]^+$ and $[CF_3COO]^-$ are formed *slowly* in nonpolar toluene on the $^{31}P$ NMR time scale, while *fast* proton–hydride exchanges still occur on the $^1H$ NMR time scale [7]. The same effect can be observed in a $CH_2Cl_2$ solution of $[ReD(PMe_3)_4(CO)]$ and $CF_3COOD$. As follows from Table 10.4, the activation energies $\Delta H^{\ddagger}$ necessary for the exchanges are notably smaller than those for the formation of dihydrogen complexes (compare with the data in Table 10.3). By contrast, the isotopic effect $k_H/k_D$ for the exchange is larger. It becomes obvious that despite the same starting point (i.e., dihydrogen-bonded complexes), the processes occur via different transition states, and thus such an exchange is not a measure of proton transfer and accompanies only the latter. States A and B in Structure 10.4 could be models of exchange transition states for the Re hydrides, where $H^+$ and $O^{2-}$ represent the acid [7]. The structures obtained by EHT calculations are not optimized but they do exhibit feasible features for the orbital interactions and explain the selectivity of the exchanges. For example, state A with a *trans* NO has a reasonable HOMO/LUMO gap ($\approx$ 2.5 eV) and crucial participation of the oxygen orbitals in the binding of the $H_2$ unit to the metal center. The orbital interactions are built up from $d_{XY}$ and $\sigma$-type functions on the Re center by the two $s_H$ hydrogen orbitals and $s$, $p_X$, and $p_Y$ orbitals from the oxygen atom. Thus, the appearance of H$\cdots$H bonding is a main feature of the exchange transition state with participation of hydride

**Structure 10.3**

**Structure 10.4**   Model of transition states A and B suitable for proton–hydride exchange processes obtained by extendend Hückel theory calculations on the basis of symmetrical pseudooctahedral geometry ($H^+$ and $O^{2-}$ represent the acid). (From ref. 7.)

ligands, in contrast to the transition state in Structure 10.3, which is typical of the usual organic bases.

### 10.3.3. Proton Transfer to a Hydride Ligand in Solution: Experimental Observation of Intermediates

The existence of stable dihydrogen-bonded complexes in solution was shown in earlier chapters. Employing the NMR spectra, Chaudret and co-workers have observed directly, for the first time, a reversible transformation of the H$\cdots$H-bonded complexes to $[\eta^2\text{-}(H_2)]^+$ systems at protonation of dihydride $RuH_2(dppm)_2$ with various OH acids [27]. Later, the location of dihydrogen bonds on the reaction coordinate of proton transfer to hydride ligands was confirmed many times by IR and NMR spectra (see reviews in refs. 25,28 and 29). For recent data, a reader can see refs. 30 and 31, where protonation of $(C_5R_5)RuH(L)$ and $TpRuH(L)$ [with R = H, Me; L = dippae, (R,R)-dippach; and Tp = hydrotris(pyrazolyl)borate] has been investigated. As we have seen, transition metal hydrides form dihydrogen bonds with average H$\cdots$H distances between 1.7 and 1.9 Å. An unusually short H$\cdots$H distance of 1.43 Å has recently been reported for the short-lived dihydrogen-bonded complex $[CpRu(PP^{Ph}PF)H\cdots HO_2CCF_3]$, which has been proposed as a new intermediate in proton transfer [32]. It should be noted that this H$\cdots$H value has been determined by $^1H$ $T_1$ time measurements in acetone-$d_6$.

Despite the intense investigations, data on H$\cdots$H bond formation rates are rather limited since the process is usually too fast for studies by standard kinetic techniques. Nevertheless, some estimations are possible from low-temperature $^1H$ NMR spectra, usually recorded in Freon solutions. These Freon solvents "stop" MH$\cdots$HX bond formation even on the $^1H$ NMR time scale. For example, the ReH resonance of $[Cp^*ReH(CO)(NO)]$ decoalesces in the presence of acidic alcohols at 96 K, giving two resolved lines at $-7.54$ and $-8.87$ ppm, assignable to the free and dihydrogen-bonded hydride, respectively [33]. Under these conditions, the lifetime of the MH$\cdots$HX complexes, $\tau$, can be calculated via

$$\tau = \frac{1}{k_1} = 2\pi \, \Delta v \tag{10.21}$$

where $\Delta v$ is the chemical shift difference for free and bonded states, expressed in hertz. The calculation gives $v \approx 10^{-3}$ s, corresponding to the rate constant, $k_1$, of about $10^4$ L/mol·s for concentrations of acidic alcohols of $< 10^{-1}$ mol/L.

Both IR and NMR techniques can be used to detect M–H···HX bonds, transforming to intimate ion pairs, $M(\eta^2-H_2)^+\cdots X^-$, and then to the product of full proton transfer. For example, Figure 10.9 illustrates the variable-temperature $^1$H NMR spectra, recorded for the hydride CpRu(CO)(PCy3)H (**1**) in CDClF$_2$/CDF$_3$ (2:1) in the presence of (CF$_3$)$_3$COH [34]. The first set of the $^1$H NMR spectra, recorded for individual hydride **1**, shows that the hydride resonance does not change when the temperature is lowered to 150 K. However, below this temperature, the line reversibly splits into two doublet resonances with an intensity ratio of 2 : 1. This behavior is observed due to the frozen motion of the cyclohexyl ring. In the presence of 12 equiv of acid (CF$_3$)$_3$COH, in addition to dihydrogen-bonded complex CpRu(CO)(PCy3)H···HOC(CF$_3$)$_3$, **3c** (it follows from the high-field shift of the hydride line), a new resonance appears in the region of $-8.0$ ppm at 160 K, which increases in intensity down to 140 K and shows a short $T_1$ time. It is obvious that the resonance belongs to the $\eta^2$-H$_2$ ligand of the dihydrogen complex, which is in equilibrium with **3c**. Below 140 K the resonance at $-8.00$ ppm broadens due to the existence of equilibrium between the contact ion pair [CpRu(CO)(PCy3)($\eta^2$-H$_2$)]$^+\cdots$[OC(CF$_3$)$_3$]$^-$, **4c**, and the ion [CpRu(CO)(PCy3)($\eta^2$-H$_2$)]$^+$, **5**, and a fast exchange between them. In the presence of 100 equivalents of (CF$_3$)$_3$COH, the $\eta^2$-H$_2$ resonance of **4c/5** appears already at 170 K. Due to the large excess of polar alcohol, increasing

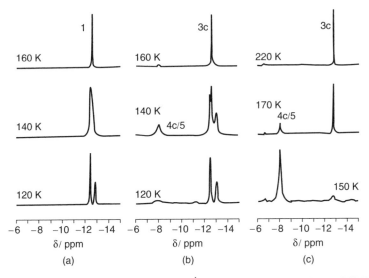

**Figure 10.9** Hydride region of the 500-MHz $^1$H NMR spectra of (1) CpRu(CO)(PCy3)H at different temperatures in CDClF$_2$/CDF$_3$ (2:1) (a) without a proton donor; (b) in the presence of 12 equiv of (CF$_3$)$_3$COH; (c) in the presence of 100 equiv of (CF$_3$)$_3$COH. (Reproduced with permission from ref. 34.)

medium polarity should lead to partial dissociation of **4c**. In fact, the resonance becomes remarkably narrower. Full proton transfer is already observed at 150 K, and at 110 K the signal of the $\eta^2$-$H_2$ ligand broadens and goes into the baseline.

The example above demonstrates the possibilities of variable-temperature NMR. However, the best experimental method for observing intimate ion pairs, $M(\eta^2 - H_2)^+ \cdots X^-$, is the IR spectroscopy. Here besides direct observation of short-living equilibrated intermediates, one can see that the positions of indicator bands [e.g., $\nu(MH)$ or $\nu(CO)$] attributable to contact ion pairs depend markedly on the nature of the proton donor HX (i.e., on hydrogen-bonded anions $X^-$) [28]. Such pairs have been found in hexane solutions of the hydride $[CpRuH(CO)(PCy_3)]$ protonated with $(CF_3)_2CHOH$, $(CF_3)_3COH$, and $CF_3COOH$ [6]. An example of time-dependent IR spectra is shown in Figure 10.2, where the assignments correspond to Scheme 10.10. The IR parameters obtained by these experiments are collected in Table 10.5, where it is seen that the $\nu(CO)$ bands of ion pairs **3** depend on HX and appear at 1972, 1978, and 2004 $cm^{-1}$ for $(CF_3)_2CHOH$, $(CF_3)_3COH$, and $CF_3COOH$, respectively. The $\nu(CO)$ frequencies of the MH$\cdots$HX complexes detected simultaneously in these solutions, also depend reasonably on $X^-$ (1931, 1935, and 1940 $cm^{-1}$), whereas the $\nu(CO)$ bands of the solvent-separated ion pairs $[CpRu(H_2)(CO)(PCy_3)]^+//CF_3COO^-$ and $[CpRu(H_2)(CO)(PCy_3)]^+//BF_4^-$ cannot be distinguished (2020 $cm^{-1}$, $CH_2Cl_2$). It is worth noting that in addition to the dihydrogen-bonded complexes **2a**, IR spectra reveal the presence of hydrogen and hydrogen/dihydrogen-bonded systems **2b** and **2c**. As was expected, increasing the solvent polarity from hexane to $CD_2Cl_2$ should lead to displacement of the positions of the equilibria in Scheme 10.10 toward contact ion pairs, due to stabilization of species with a larger charge separation. In accordance with this statement, the IR spectra of $[CpRuH(CO)(PCy_3)]$ in $CH_2Cl_2$ solutions and in the presence of $(CF_3)_3COH$ exhibit $\nu(CO)$ bands belonging only to contact ion pairs. In addition, the $[CpRuH(CO)(PCy_3)]/CF_3COOH$ system can show only intimate ion pairs, only solvent-separated ion pairs, or both species simultaneously as a function of the $CF_3COOH$/hydride ratio. Finally, the data show that contact ion pairs are energetically preferable with respect to dihydrogen-bonded complexes, particularly in more polar solvents such as $CH_2Cl_2$.

## 10.4. ENERGY PROFILE OF PROTON TRANSFER TO A HYDRIDE LIGAND IN SOLUTION

The spectroscopic, kinetic, and thermodynamic data discussed are sufficient to describe semiquantitatively the energy profile of proton transfer to a hydride ligand occurring in solution [29, 35, 36]. Figure 10.10 shows the energy as a function of the proton–hydride distance, varying from the initial state to a final product. The average structural parameters of the initial hydrides and intermediates have been taken from earlier chapters. Since proton–hydride contacts of

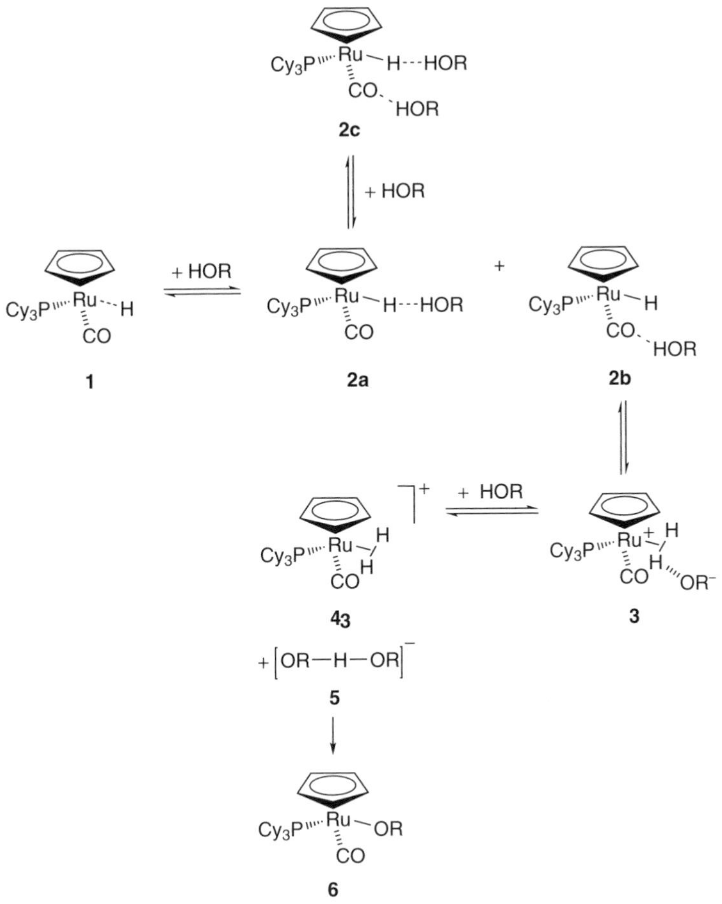

$$Cy_3P^{\text{\tiny\textbackslash}}\text{-}\overset{\displaystyle\bigcirc}{Ru}\text{-}H\cdots HOR$$
$$\underset{\phantom{CO}}{CO_{\text{\tiny\textbackslash}}}\quad\phantom{xx}$$
$$\phantom{xxxx}\text{'}HOR$$

**2c**

$+ HOR$

$$Cy_3P^{\text{\tiny\textbackslash}}\text{-}\overset{\displaystyle\bigcirc}{Ru}\cdots H \qquad \overset{+\,HOR}{\rightleftharpoons} \qquad Cy_3P^{\text{\tiny\textbackslash}}\text{-}\overset{\displaystyle\bigcirc}{Ru}\text{-}H\cdots HOR \qquad + \qquad Cy_3P^{\text{\tiny\textbackslash}}\text{-}\overset{\displaystyle\bigcirc}{Ru}\text{-}H$$
$$\underset{CO}{\phantom{x}} \qquad\qquad\qquad \underset{CO}{\phantom{x}} \qquad\qquad\qquad \underset{CO_{\text{\tiny\textbackslash}}}{\phantom{x}}$$
$$\phantom{xxxxxxxxxxxxxxxxxxxxxxxxxxxxxxxxxx}\text{'}HOR$$

**1**        **2a**        **2b**

$$\left[Cy_3P^{\text{\tiny\textbackslash}}\text{-}\overset{\displaystyle\bigcirc}{Ru}\overset{H}{\underset{H}{\diagdown}}\right]^+ \quad \overset{+\,HOR}{\rightleftharpoons} \quad Cy_3P^{\text{\tiny\textbackslash}}\text{-}\overset{\displaystyle\bigcirc}{Ru}^{\pm}\overset{H}{\underset{H_{\text{\tiny\textbackslash}}}{\diagup}}$$
$$\underset{CO}{\phantom{x}} \qquad\qquad\qquad\qquad \underset{CO}{\phantom{x}}\quad\phantom{x}OR^-$$

**43**          **3**

$$+\left[OR\text{-}H\text{-}OR\right]^-$$

**5**

$$Cy_3P^{\text{\tiny\textbackslash}}\text{-}\overset{\displaystyle\bigcirc}{Ru}\text{-}OR$$
$$\underset{CO}{\phantom{x}}$$

**6**

**Scheme 10.10**

$\leq 2.4$ Å found for different compounds in the solid state correspond to the appearance of weak bonding interactions, the $H\cdots H$ distance of 2.5 Å can be regarded as the starting point with zero energy. The proton transfer process ends with a structure in which the $H^{\delta+}\cdots^{\delta-}H$ distance becomes 0.9 Å, a typical value for dihydrogen ligands.

The pattern in Figure 10.10 contains four energy minima occupied by the initial hydride, the dihydrogen-bonded complex, the intimate ion pair, and the solvent-separated ion pair as a final product of the reaction. For simplicity, the homo-conjugated species $[RO\cdots H\cdots OR]^-$, assisting proton transfer and stabilizing the protonated products, can be ruled out.

Generally, the formation of hydrogen-bonded complexes is regarded as a very fast, diffusion-controlled process with no energy barrier. On the other hand, even ultrafast molecular reorientations in solution require energy for reorganization of

**TABLE 10.5. IR Spectral Parameters for the Compounds in Scheme 10.10** [a]

| System | $v(CO)(cm^{-1})$ in Complex: | | | | | Solvent |
|---|---|---|---|---|---|---|
| | 1 | 2a | 3 | 4 | 6 | |
| **1** | 1920 | — | — | — | — | Hexane |
| **1** + (CF$_3$)$_2$CHOH | 1920 | 1931 | 1972 | — | 1946 | Hexane |
| **1** + (CF$_3$)$_3$COH | 1920 | 1935 | 1978 | — | 1955 | Hexane |
| **1** + CF$_3$COOH | 1920 | 1940 | 2004 | — | 1963 | Hexane |
| **1** + (CF$_3$)$_3$COH | 1890 | — | 1960 | — | 1944 | CH$_2$Cl$_2$ |
| **1** + (CF$_3$)$_3$COH | 1890 | — | 1996 | 2020 | 1957 | CH$_2$Cl$_2$ |
| **1** + HBF$_4$ | — | — | — | 2020 | — | CH$_2$Cl$_2$ |

*Source:* Ref. 6.
[a]Complex **1**, initial hydride; **2a**, its dihydrogen-bonded complex; **3**, contact ion pair; **4**, solvent-separated ion pair. Compounds **4** plus **5** are transformed to complex **6** after H$_2$ elimination.

molecular environments. Since transition metal complexes reorient in solutions with energies of 3.0 to 4.5 kcal/mol, this magnitude can be taken as a reasonable first barrier, separating the free hydride molecules and the H···H complexes. Dihydrogen bonding leads to systems with H···H separations between 1.6 and 1.9 Å. As we have shown previously, the bonding energy in these systems is reduced by 2.0 to 7.6 kcal/mol and depends to a significant degree on the nature of hydrides and acids, whereas the solvent influence is insignificant. The M–H and H–O bonds in the dihydrogen-bonded complex are only slightly elongated with respect to the initial hydride and the acid. The electronic perturbations are minimal and do not require significant energy. The next energy minimum is populated by contact ion pairs that are stabilized by hydrogen bonds with anions. This intermediate has H···H separation very close to that in the dihydrogen complex, and its H···O bond is strongly elongated. Despite significant structural changes, the depths of the energy minima occupied by dihydrogen-bonded complexes and contact ion pairs are expected to be similar. Since the ion-pair species were not localized in the gas-phase DFT calculations (see below), the effect is certainly consistent with their stabilization by polar solvents or homo-conjugated anions [RO···H···OR]⁻. As in the case of classical acid–base pairs, the formation of contact ion pairs $M(H_2)^+$···X⁻ is the step of proton transfer. However, for transition metal hydrides, this step requires significant changes in the M–H···H–X geometry and the H···H and X–H bond lengths. For this reason, the energy barrier, which separates the M–H···H–X and $M(H_2)^+$···X⁻ species, increases to 11 to 13 kcal/mol. It is obvious that this step is responsible for the total kinetics of proton transfer and the large variation in the $k$ constants in Table 10.1. The solvent-separated ion pair finishes the process with enthalpy changes between − 2.4 and − 9 kcal/mol [25, 36]. NMR experiments on the [CpRuH(CO)(PCy$_3$)]/(CF$_3$)$_3$–COH system in CDClF$_2$/CDF$_3$ (2 : 1) [33] showed a low-energy barrier that separates the contact and solvent-separated ion

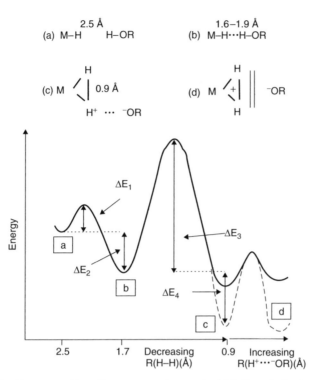

**Figure 10.10** Energy profiles of proton transfer to a hydride ligand of a transition metal complex in solution: $\Delta E_1 = +3$ to $4\,\text{kcal/mol}$, $\Delta E_2 = -5$ to $-7\,\text{kcal/mol}$, $\Delta E_3 = +10$ to $14\,\text{kcal/mol}$, and $\Delta E_4 = -7\,\text{kcal/mol}$; the energy is a function of the proton–hydride distance, varying from an initial state (2.5 Å) to the final product (0.9 Å); conversion of the intimate ion pair to the solvent-separated ion pair is shown as a function of the $H^+\cdots O^-$ distance. (Reproduced with permission from ref. 29.)

pairs. Therefore, in the absence of quantitative kinetic data on the last step, the barrier between $M(H_2)^+\cdots X^-$ and $M(H_2)^+//X^-$ can be taken to be small.

Similar kinetic and thermodynamic data, obtained for transformations of $OsH_2(PP_3)$ to the dihydrogen complex $[OsH(H_2)(PP_3)]^+$ [40] and for $Cp^*Ru(dppe)H$ to the $[Cp^*Ru(dppe)(H_2)]^+$ [41] in the presence of various acids, have resulted in the same energy profile of the reactions.

## 10.5. ENERGY, INTERMEDIATES, AND TRANSITION STATES IN INTERMOLECULAR PROTON TRANSFER TO HYDRIDIC HYDROGENS: THEORETICAL STUDIES

Modern computational chemistry provides a detailed analysis of the potential energy surface of reactants when they are transformed to products of a reaction. This analysis, the most important advantage available in the theoretical approaches, allows us to trace a minimum energy path from reactants to products via intermediates and transition states. In turn, the intermediates and transition states can also be characterized in detail. It is obvious that such characterizations help to reveal the factors reducing energy barriers on reaction coordinates, and thus to control the kinetics of the transformations. In the framework of proton transfer to hydridic hydrogens, analysis of the potential energy surface is directed to localization of structurally different dihydrogen-bonded complexes, contact and separated ionic pairs, and transition states lying between them. In addition, potential energy surface studies can show how the reaction path changes in going from the gas phase to the condensed state, due to the appearance of different stabilizing interactions.

Despite the large variety of theoretical approaches and levels of calculation, the methodology of potential-energy-surface investigations includes determination of all stationary points and vibrational analysis showing their nature. Then a relaxed potential-energy-surface scan is reached by stepwise shortening of the proton–hydride distance. It was found that the highest-energy structure can then be optimized to locate a transition state. Finally, it should be emphasized that in contrast to experimentalists, theoreticians have more freedom in choosing objects of investigation. The latter is particularly important for revealing the factors connected with the influence on the nature of elements of hydrides that accept protons. For all the relevant data, readers are referred to refs. 25,28,35, and 37. Here, we consider only examples of the theoretical studies most similar to systems in the context of the kinetic investigations described above.

### 10.5.1. Proton Transfer to Hydrides of the Group 3A Elements

Marincean and Jackson [38] have performed MP2//6-311++G** calculations for proton transfer from acids $H_2O$, HF, and HCl to ions $[AlH_4]^-$ to form covalent-bonded products after $H_2$ elimination:

$$[AlH_4]^- + HX \rightarrow [AlH_3X]^- + H_2 \qquad (10.22)$$

According to the calculations, the reagents form the dihydrogen-bonded complexes, shown in Figure 10.11, with typical geometry and H$\cdots$H distances of less than 2.4 Å. Only in the case of the complex $[AlH_4 \cdots H_2O]^-$ do both protons of the water molecule participate in dihydrogen bonding, leading to $C_{2v}$ symmetry of the complex. In each case the potential energy surface has shown only one maximum corresponding to a transition state. As shown in the figure, all the transition states exhibit proton and hydride atoms that are identically located with respect to the Al center, with distances $\approx$ of 2.0 Å. Thus, in the gas phase, exothermic proton transfer occurs somewhat in the framework of a concerted mechanism. Figure 10.12 illustrates this process for the $[AlH_4 \cdots HCl]^-$ complex, where a proton transfer energy barrier is equal to 6.7 kcal/mol.

Theoretical studies of proton transfer from proton donors of different strength, $CH_3OH$, $CF_3CH_2OH$, and $CF_3OH$, to hydridic hydrogens of the ions $[AlH_4]^-$, $[GaH_4]^-$, and $[BH_4]^-$ have been carried out by the DFT/B3LYP and MP2

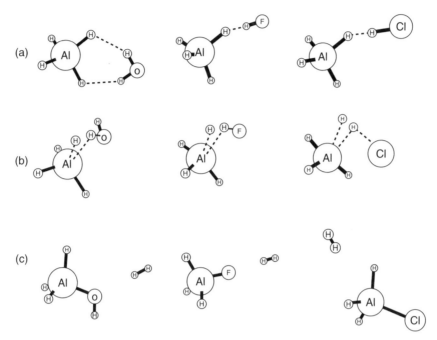

**Figure 10.11** MP2//6-311++G** calculations of (a) the reagents, (b) transition states, and (c) final products for proton transfer according to eq. (10.22). (Reproduced with permission from ref. 38.)

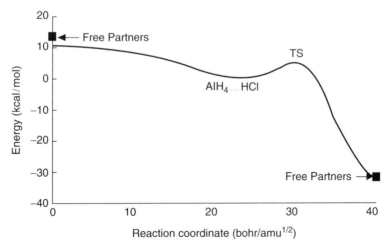

**Figure 10.12** Energy profile obtained for proton transfer from HCl via dihydrogen-bonded complex [AlH$_4$···HCl]$^-$ along the reaction coordinate. (Reproduced with permission from ref. 38.)

methods using the 6-311 ++G(d,p) basis set [39]. The two methods show similar results where dihydrogen-bonded complexes are lying on a reaction coordinate. In contrast to transition metal hydride systems, the unstable $\eta^2$-$H_2$ complexes of Al, Ga, and B play the role of *transition states* (but not intermediates), which are also stabilized by a hydrogen bond with the $OR^-$ anion. Again, the process has a rather concerted character in which the proton transfer, $H_2$ release, and formation of covalent-bonded products occur in a single step. At the same time, another mechanism and energy profile have been found for the $BH_4^-$/$CF_3OH$ system and are shown in Figure 10.13. In this case, a $(\eta^2$-$H_2)B$ complex, stabilized by a hydrogen bond, exists as an intermediate, separated from **TS1** by a very low energy barrier. The solvent influence has been probed in this work in the framework of a polarizable conductor calculation model, simulating the presence of THF for proton transfer from $CF_3CH_2OH$ (Figure 10.14). It follows from the data that THF leads to remarkable *destabilization* of the dihydrogen-bonded intermediates and transition states relative to the reagents. The products of the reaction are also destabilized. At the same time, the energy barriers for the proton transfer remain similar and only slightly affected by the solvent.

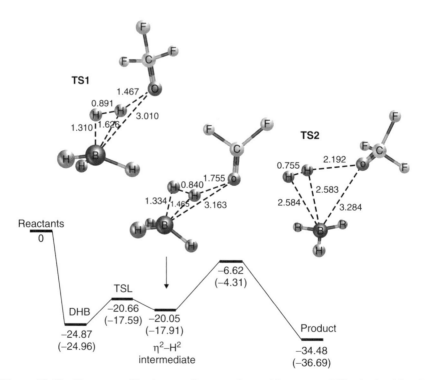

**Figure 10.13** Energy profile, intermediates, and transition states (TS) obtained by the B3LYP and MP2 (in parentheses) methods for proton transfer from $CF_3OH$ to the hydridic hydrogen of the ion $[BH_4]^-$. (Reproduced with permission from ref. 39.)

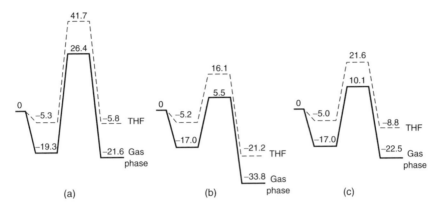

**Figure 10.14** Energy profiles for the reaction of (a) $[BH_4]^-$, (b) $[AlH_4]^-$, and (c) $[GaH_4]^-$ with $CF_3CH_2OH$ in the gas phase and in THF. (Reproduced with permission from ref. 39.)

### 10.5.2. Theory of Proton Transfer to Transition Metal Hydrides

Orlova and Scheiner [42] have investigated at the DFT/ B3PW91 level the gas-phase protonation of the transition metal hydrides $(Cp)Re(H)(NO)(CO)$, $(Cp)Ru(H)(CO)(PH_3)$, and $(Cp)Re(H)(NO)(PH_3)$ with poor, moderate, and strong proton donors $H_2O$, $HOCF_3$, and $H_3O^+$. As in the case of hydrides of the group 3A elements, it has been found that reaction pathways depend on the proton- and proton-donor strengths. For example, the hydride $(Cp)Re(H)(NO)(CO)$ forms dihydrogen-bonded intermediates with weak and strong proton donors $H_2O$ and $H_3O^+$ that transform to the $\eta^2$-$H_2$ complex due to proton transfer over a very small energy barrier. The hydride $(Cp)Ru(H)(CO)(PH_3)$, which has a highly nucleophilic hydride ligand, gives a dihydrogen-bonded intermediate with moderate proton donor $HOCF_3$, while the stronger donor $H_3O^+$ leads to the $\eta^2$-$H_2$ complex only. Finally, it has been established that in the hydride $(Cp)Re(H)(NO)(PH_3)$, a preferable proton-accepting site is the metal atom.

Despite reliable experiments with observations of contact ion pairs, gas-phase DFT calculations [6, 43] of the ion pair structures $[CpRu(CO)(PH_3)(H_2)]^+\cdots$ $OC(CF_3)_3^-$ and $[CpRu(CO)(PH_3)(H_2)]^+\cdots OCOCF_3^-$ or $[Cp_2NbH_2-(H_2)]^+\cdots$ $OC(CF_3)_3^-$ and $[Cp_2NbH_2(H_2)]^+\cdots OCOCF_3^-$ have ended up in the initial dihydrogen-bonded complexes. In addition, the products of full proton transfer (i.e., free ions) were thermodynamically unstable and their energies were higher than those of the reagents by 100 kcal/mol. In this connection, let us consider in detail the protonation of the hydride $CpRu(CO)(PH_3)H$ with $H_3O^+$, $CF_3COOH$ (**TFA**), and $(CF_3)_3COH$ (**PFTB**), investigated by DFT/B3LYP calculations [6].

When the initial hydride $CpRuH(CO)(PH_3)$, **1t**, is placed near $H_3O^+$, a proton transfer occurs without the energy barrier to form the dihydrogen complex $[CpRu(H_2)(CO)(PH_3)]^+$, in good agreement with the data of Orlova and Scheiner [42]. Dihydrogen-bonded intermediates appear on the protonation pathway if the hydride **1t** interacts with the weaker acids TFA and PFTB. However, the search for the ionic pairs [CpRu (CO)(PH$_3$)(H$_2$)]$^+\cdots$OR$^-$ has led to negative results in the gas phase. Solvent effects simulating the presence of heptane and CH$_2$Cl$_2$ have revealed their strong influence on the thermodynamics of charged species. Nevertheless, even for strong proton donor TFA and the relatively polar dichloromethane, contact ion pairs were not observed and the solvent-separated ions remained too energetically rich; they were lying higher than the initial reagents by 22 kcal/mol. At the same time, the contact ion pair intermediates have been located with minimal energies even in the gas phase, when homo-conjugated anions, [RO$\cdots$H$\cdots$OR]$^-$, have been taken into consideration as an additional factor for charge stabilization. The calculations have resulted in the structure [CpRu(CO)(PH$_3$)(H$_2$)]$^+\cdots$[RO$\cdots$H$\cdots$OR]$^-$ with the optimized geometry shown in Structure 10.5. As can be seen, one of the dihydrogen atoms in [CpRu(CO)(PH$_3$)(H$_2$)]$^+\cdots$[CF$_3$C(O)–O$\cdots$H$\cdots$O(O)CCF$_3$]$^-$ acts as a proton donor and forms a hydrogen bond of length 1.891 Å. The corresponding Ru–H distance (1.820 Å) is slightly longer with respect to the free cationic dihydrogen complex (1.765 and 1.775 Å), while the H$\cdots$H separations in the H$_2$ ligands are very similar in both species (0.847 Å). Finally, the H–O bond in the ion pair involved in hydrogen bonding is elongated significantly with respect to that in the free proton donor (1.253 and 0.98 Å, respectively).

The critical role of the solvent polarity in the stabilization of the hydrogen-bonded ion pairs and the transition states in the charge separation steps is illus-trated in Figure 10.15, where **2t-2TFA** is the dihydrogen-bonded complex and **3t-2TFA** is the contact ion pair, [CpRu(CO)(PH$_3$)(H$_2$)]$^+\cdots$[CF$_3$C(O)O$\cdots$H$\cdots$O(O)–CCF$_3$]$^-$. Thus, in good agreement with the kinetic data, the last step shows a maximal energy barrier. Finally, Figure 10.16 allows us to trace how the energy varies at the proton movement, which transforms the dihydrogen-bonded complex into the ionic pair.

The unusual kinetics observed for proton transfer to a hydride ligand in the cluster [W$_3$S$_4$(PH$_3$)$_6$H$_3$]$^+$ has been interpreted by a mechanism that includes two

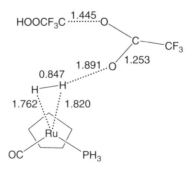

**Structure 10.5** DFT-optimized geometry of the hydrogen-bonded contact ion pair [CpRu(CO)(PH$_3$)(H$_2$)]$^+\cdots$[CF$_3$C(O)O$\cdots$H$\cdots$O(O)–CCF$_3$]$^-$ stabilized by the [CF$_3$C(O)O$\cdots$H$\cdots$O(O)CCF$_3$]$^-$ anion in the gas phase. The distances are given in angstroms.

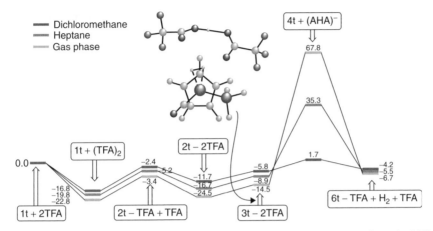

**Figure 10.15** Energy profile obtained by DFT calculations for proton transfer to hydridic hydrogen in hydride **1t** from the dimer $(CF_3COOH)_2$ (**TFA**). Energies are given in kcal/mol. (Reproduced with permission from ref. 6.)

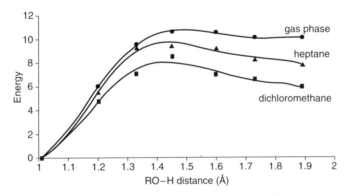

**Figure 10.16** Potential energy curves for the transformation of dihydrogen-bonded complex **2t-2TFA** to contact ion pair **3t-2TFA**. The O–H distance is taken as a reaction coordinate of the transferring proton. Energies are given in kcal/mol. (Reproduced with permission from ref. 6.)

competitive pathways [9]:

$$W\text{–}H + HCl \rightleftharpoons W\text{–}H \cdots H\text{–}Cl \rightleftharpoons [TS1] \rightarrow W\text{–}Cl + H_2$$

$$W\text{–}H + 2HCl \rightleftharpoons W\text{–}H \cdots H\text{–}Cl \cdots H\text{–}Cl \rightleftharpoons [TS2] \rightarrow W\text{–}Cl \cdots H\text{–}Cl + H_2$$

(10.23)

To support this mechanism independently, the authors have performed DFT/ B3LYP calculations with full optimizations of geometries and energies for both

**TABLE 10.6. Relative Energies (kcal/mol) Calculated for Intermediates and Transition States TS1 and TS2**[a]

| Structure | Gas Phase | $CH_2Cl_2$ | $CH_3CN$ |
|---|---|---|---|
| $W-H\cdots H-Cl$ | $-3.07$ | $-2.90$ | $-2.79$ |
| TS1 | $12.07$ | $10.48$ | $10.30$ |
| $W-Cl + H_2$ | $-28.23$ | $-27.36$ | $-27.18$ |
| $W-H\cdots H-Cl\cdots H-Cl$ | $-9.21$ | $-7.74$ | $-7.45$ |
| TS2 | $5.42$ | $4.49$ | $4.44$ |
| $W-Cl\cdots H-Cl + H_2$ | $-32.05$ | $-30.61$ | $-30.33$ |

[a]These states lie on the reaction coordinate of proton transfer to a hydride ligand of cluster $[W_3S_4(PH_3)_6H_3]^+$ according to eq. (10.23).

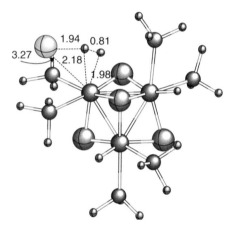

**Figure 10.17**  Optimized geometry of the transition state TS1 calculated for proton transfer via dihydrogen-bonded complex $W-H\cdots H-Cl$. The distances are given in angstroms. (Reproduced with permission from ref. 9.)

intermediates and transition states. The relative energies are shown in Table 10.6, and the structures of transition states **TS1** and **TS2** are illustrated in Figures 10.17 and 10.18. As is shown clearly, the $H\cdots H$ distance in transition state **TS1** is already short (0.81 Å) and the two $W\cdots H$ distances are determined as 1.98 and 2.18 Å. In other words, this structure represents a dihydrogen complex that is ion-paired with the ion $Cl^-$. This complex then eliminates the dihydrogen molecule. The more energetically preferable transition state, **TS2** (Table 10.6), exhibits a shorter $H\cdots H$ distance of 0.78 Å, while the $W\cdots H$ distances (2.10 and 2.25 Å) are slightly elongated with respect to **TS1**. Again, the transition state is a dihydrogen complex that is ion-paired with the anion $HCl_2^-$. Thus, in contrast to the $CpRu(CO)(PH_3)H$ system considered above, ion pairs stabilized by hydrogen bonds with one or two acid molecules appear on the reaction coordinate

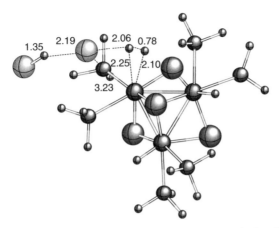

**Figure 10.18** Optimized geometry of the transition state TS2 calculated for proton trans-fer via the hydrogen/dihydrogen-bonded complex W–H···H–Cl···H–Cl. The distances are given in angstroms. (Reproduced with permission from ref. 9.)

as transition states but not as intermediates. Table 10.6 shows that both transition states (**TS1** and **TS2**) can be stabilized by solvents $CH_2Cl_2$ and $CH_3CN$ with respect to the gas phase. However, the effects are not dramatic. Finally, since the total energy barriers obtained are similar for the two reaction pathways, an excess of the acid can actually increase the probability of the second pathway, in good agreement with the experimental kinetic data discussed earlier.

## 10.6. CONCLUDING REMARKS

1. The mechanism of proton transfer to hydridic hydrogens depends strongly on the strengths of proton-donor and proton-acceptor sites, the nature of the element, bonding to a hydridic hydrogen, and also on the aggregate states (the solid state, the gas phase, or solution) and medium polarity.

2. Despite these circumstances, one can formulate a general mechanism of proton transfer. It starts with dihydrogen bonding as the preorganizing interaction, occurs via H···H complexes and hydrogen-bonded contact ion pairs, and ends in dihydrogen complexes that can eliminate $H_2$. The dihydrogen-bonded complexes and ion pairs, appearing as intermediates or transition states, can be stabilized by the second acid molecule. Such $[X···H···X]^-$ assistance can dictate the total kinetic order of the process.

3. General rules governing proton transfer to hydride ligands and conventional organic bases are very similar. The difference between them is quantitative rather than qualitative in character. In the case of proton transfer to hydridic hydrogens, the contact ion pair formation step, which determines the kinetics of the process, requires significantly larger activation energies, and for this reason the total rate of proton transfer is remarkably slower.

## REFERENCES

1. M. Peruzzini, R. Poli, *Recent Advances in Hydride Chemistry*, Elsevier, Amsterdam (2001).
2. X. Zhao, I. P. Goergakaki, M. I. Miller, I. C. Yarbourh, M. Y. Darensbourg, *J. Am. Chem. Soc.* (2001), **123**, 9710.
3. J. W. Peters, *Curr. Opin. Struct. Biol.* (1999), **9**, 670.
4. M. A. Esteruelas, L. A. Oro, *Chem. Rev.* (1998), **98**, 577.
5. M. G. Basalotte, J. Duran, M. J. Fernandez-Trujillo, M. A. Manez, J. R. de la Torre, *J. Chem. Soc. Dalton Trans.* (1998), 745.
6. N. V. Belkova, M. Besore, L. M. Epstein, A. Lledos, F. Maseras, E. S. Shubina, *J. Am. Chem. Soc.* (2003), **125**, 7715.
7. S. Feracin, T. Burgi, V. I. Bakhmutov, I. Eremenko, E. V. Vorontsov, A. B. Vimenits, H. Berke, *Organometallics* (1994), **13**, 4194.
8. R. K. Harris, *Nuclear Magnetic Resonance Spectroscopy: A Physicochemical View*, Longman Scientific & Technical, Harlow, England (1986), pp. 119–130.
9. A. G. Algarra, M. G. Basallote, M. Feliz, M. J. Fernandez-Trujillo, R. Llusar, V. S. Safont, *Chem. Eur. J.* (2006), **12**, 1413.
10. T. Zieger, J. Autschbach, *Chem. Rev.* (2005), **105**, 2695.
11. I. Rohaz, I. Alkorta, J. Elguero, *Chem. Phys. Lett.* (1997), **275**, 423.
12. J. W. Hwang, J. P. Cambell, J. Kozubowski, S. A. Hanson, J. F. Evants, W. L. Gladfelter, *Chem. Mater.* (1995), **7**, 517.
13. R. Custelcean, J. E. Jackson, *J. Am. Chem. Soc.* (1998), **120**, 12935.
14. R. Custelcean, J. E. Jackson, *J. Am. Chem. Soc.* (2000), **122**, 5251.
15. A. J. Lough, S. Park, R. Ramachandar, R. H. Morris, *J. Am. Chem. Soc.* (1994), **116**, 8356.
16. G. J. Kubas, *Metal Dihydrogen and s-Bond Complexes*, Kluwer Academic/Plenum Press, New York (2001).
17. R. H. Morris, in *Recent Advances in Hydride Chemistry*, ed. R. Poli and M. Peruzzini, Elsevier, New York (2001), pp. 1–38.
18. E. T. Papich, F. C. Rix, N. Spetseris, J. R. Norton, R. D. Williams, *J. Am. Chem. Soc.* (2000), **122**, 12235.
19. M. G. Basallote, J. Duran, M. J. Fernandez-Trujillo, M. A. Manez, *J. Chem. Soc. Dalton Trans.* (1998), 2205.
20. M. G. Basallote, J. Duran, M. J. Fernandez-Trujillo, M. A. Manez, *Organometallics* (2000), **19**, 695.
21. M. G. Basallote, J. Duran, M. J. Fernandez-Trujillo, M. A. Manez, *Inorg. Chem.* (1999), **38**, 5067.
22. N. V. Belkova, P. O. Revin, L. Epstein, E. V. Vorontsov, V. I. Bakhmutov, E. S. Shubina, E. Collange, R. Poli, *J. Am. Chem. Soc.* (2003), **125**, 11106.
23. N. S. Golubev, S. N. Smirnov, V. A. Gindin, G. S. Denisov, H. Benedict, H. H. Limbach, *J. Am. Chem. Soc.* (1994), **116**, 12055.
24. P. Schah-Mohammedi, I. G. Shenderovich, C. Detering, H. H. Limbach, P. M. Tolstoy, S. N. Smirnov, G. S. Denisov, N. S. Golubev, *J. Am. Chem. Soc.* (2000), **122**, 128787.

25. N. V. Belkova, E. I. Gutsul, E. S. Shubina, L. M. Epstein, *Z. Phys. Chem.* (2003), **217**, 1525.

26. N. S. Golubev, G. S. Denisov, *Zh. Prikl. Spectrosc.* (1982), **37**, 265.

27. J. A. Ayllon, S. F. Sayers, S. Sabo-Etienne, B. Donnadieu, B. Chaudret, E. Clot, *Organometallics* (1999), **18**, 3981.

28. L. M. Epstein, E. S. Shubina, *Coord. Chem. Rev.* (2002), **231**, 165.

29. V. I. Bakhmutov, *Eur. J. Inorg. Chem.* (2005), 245.

30. M. Jimenez-Tenorio, M. Dolores Palacios, M. Carmen Puerta, P. Valegra, *Inorg. Chem.* (2007), **46**, 1001.

31. M. Jimenez-Tenorio, M. Dolores Palacios, M. Carmen Puerta, P. Valegra, *Organometallics* (2006), **25**, 4019.

32. E. Cayuela, F. A. Jalon, B. R. Manzano, G. Espino, W. Weissensteiner, K. Mereiter, *J. Am. Chem. Soc.* (2004), **126**, 7049.

33. E. S. Shubina, N. V. Belkova, A. V. Ionidis, N. S. Golubev, L. M. Epstein, *Russ. Chem. Bull.* (1997), **44**, 1349.

34. N. V. Belkova, A. V. Ionidis, L. M. Epstein, E. S. Shubina, S. Gruendemann, N. S. Golubev, H. H. Limbach, *Eur. J. Inorg. Chem.* (2001), 1753.

35. L. M. Epstein, N. V. Belkova, E. S. Shubina, in *Recent Advances in Hydride Chemistry*, ed. R. Poli and M. Peruzzini, Elsevier, New York (2001), pp. 391–418.

36. N. V. Belkova, E. V. Bakhmutova, E. S. Shubina, C. Bianchini, M. Peruzzini, V. I. Bakhmutov, L. M. Epstein, *Eur. J. Inorg. Chem.* (2001), 2163.

37. R. Custelcean, J. E. Jackson, *Chem. Rev.* (2001), **101**, 1963.

38. S. Marincean, J. E. Jackson, *J. Phys. Chem. A* (2004), **108**, 5521.

39. O. Filippov, A. M. Filin, V. N. Tsupreva, N. V. Belkova, A. Lledos, G. Ujaque, L. M. Epstein, E. S. Shubina, *Inorg. Chem.* (2006), **45**, 3086.

40. E. I. Gutsul, N. V. Belkova, M. S. Sverdlov, E. S. Shubina, L. Epstein, V. I. Bakhmutov, M. Perruzzini, C. Binachini, F. Zanobini, T. N. Gribanova, R. M. Minyaev, *Chem. Eur. J.* (2003), **9**, 2219.

41. N. V. Belkova, P. A. Dub, M. Bata, J. Houghton, *Inorg. Chim. Acta* (2007), **360**, 149.

42. G. Orlova, S. Scheiner, *J. Phys. Chem.* (1998), **102**, 4813.

43. E. V. Bakhmutova, V. I. Bakhmutov, N. V. Belkova, M. Besora, L. Epstein, A. Lledos, G. I. Nikonov, E. S. Shubina, J. Tomas, E. V. Vorontsov, *Chem. Eur. J.* (2004), **10**, 661.

# 11

# GENERAL CONCLUSIONS

Numerous experimental and theoretical data discussed herein prove the existence of dihydrogen bonds in the gas phase, in solution, and in the solid state and formulate the common concept of dihydrogen bonding. Each chapter has ended with a set of concluding remarks. In these sections a reader can find details concerning intra- and intermolecular dihydrogen bonds, the nature of the elements donating hydridic hydrogens, the role of dihydrogen bonds in molecular aggregations, stabilization of molecular conformations, and proton transfer reactions, where dihydrogen-bonded complexes can appear as intermediates or transition states. In the present chapter we sum up the most general conclusions and focus on two general questions: How short or long can a dihydrogen bond be, and what environmental factors act particularly strongly on dihydrogen bonding?

It is now accepted that H$\cdots$H contacts smaller than 2.4 Å between two hydrogen atoms that have opposite charges can be termed dihydrogen bonds. Similarly to classical hydrogen bonds, the bonding energy of dihydrogen bonds is dominated by electrostatic interaction. In this sense, what makes a dihydrogen bond so unusual is that the proton-accepting site is also a hydrogen atom. Nevertheless, one can define the following important difference between these types of bonding. If the electrostatic component in regular hydrogen bonds is followed by charge transfer energy $E_{CT}$ and the polarization contribution $E_{PL}$ is very small, dihydrogen bonds contain a much larger $E_{PL}$ component and thus can be termed $E_{ES} > E_{CT} \approx E_{PL}$. In some cases, the $E_{CT}$ contribution can even dominate.

## 11.1. HOW SHORT OR LONG CAN DIHYDROGEN BONDS BE?

The H$\cdots$H distances in dihydrogen-bonded complexes change in a very large region as a function of proton-donor and proton-acceptor strengths. At weak polarization of initial bonds, donating protons and hydridic hydrogens, dihydrogen bonds can be strongly elongated. Typical examples are complexes related to

*Dihydrogen Bonds: Principles, Experiments, and Applications*, By Vladimir I. Bakhmutov
Copyright © 2008 John Wiley & Sons, Inc.

$BH_4^-\cdots H_4C$, where despite the pronounced hydridic character of ions $BH_4^-$, the proton donor strength of bond C–H is extremely weak, leading to H···H distances even longer than 2.4 Å (2.6 to 2.8 Å) and bonding energies around 1 kcal/mol. In accord, the electron density in the H···H directions is also very small. In these terms, weak complexes are closer to van der Waals systems. Nevertheless, the main attribute of dihydrogen bonding, directionality, remains. On the other hand, since the energy controlling this directionally is insignificant, it becomes clear that there is no clear border between weak dihydrogen bonds and van der Waals interactions.

H···H complexes formed in the absence of electrostatic attraction represent the special class of interaction that Bader called H–H bonding. This bonding takes place when both of the hydrogen atoms show practically zero charges (e.g., C–H···H–C) or the atomic charges are identical (e.g., M–H···H–M). In the last case, the nature of the bonding is not simple: The weak Mn–H···H–Mn contacts represent a four-electron destabilizing interaction between the $\sigma$ M–H bonding levels where mixing of the $\sigma^*$ M–H levels produces a weakly bonding character. The H···H distances in the foregoing systems are again very long and close to those in van der Waals complexes.

The other end of the distance scale corresponds to the strongest dihydrogen bonds, which are particularly interesting in the context of rapid self-organization of molecular building blocks into extended regular structures or fast proton transfer along these bonds. For example, in transition metal hydride systems, short-lived dihydrogen-bonded intermediates with unusually small H···H distances (1.4 to 1.5 Å) have been localized on a reaction coordinate of proton transfer to a hydrogen ligand (see Chapter 10).

The shortest dihydrogen bonds have recently been analyzed by Grabowski and co-workers [1–4] as a separate important topic, considering this type of bonding as a part of covalent chemical bonds. Earlier it was found that classical hydrogen bonds H···Y have a partial covalent character when they are very strong and accompanied by resonance or charge assistance [1]. As in the case of regular chemical bonds, these hydrogen bonds show the *negative* Laplacian of the electron density at the H···Y bond critical points. The same criterion has been applied for the strong charge-assisted beryllium dihydrogen-bonded complexes [2–4]. According to ab initio calculations performed at different theoretical levels, the beryllium complexes do not exhibit imaginary frequencies and thus occupy energy minima on the potential energy surface. Some important structural and electron density parameters of the complexes and their bonding energies obtained at the MP2/6-311++G(d,p) level are listed in Table 11.1.

Figure 11.1 illustrates the maxima of the electron density observed in the relief map, computed for the charge-assisted complex $H_2OH^+\cdots H–Be–Be–H$. The figure is lying in the plane going via the atoms of the accepting molecule H–Be–Be–H and the proton donor bond O–H. The maxima of two other hydrogen atoms in the proton-donating molecule $H_3O^+$ are not visible because their attractors are lying below the plane of the figure. It is easy to see that these dihydrogen-bonded complexes exhibit extremely short H···H contacts between

**TABLE 11.1. H···H, O–H, and Be–H Distances, Electron Density Parameters, and BSSE-Corrected Bonding Energies of Strong Dihydrogen-Bonded Complexes$_a$**

| Complex | $-\Delta E$ (kcal/mol) | H···H (Å) | $\Delta r$(O–H) (Å) | $\Delta$(Be–H) (Å) | $\rho_C$ (au) | $\nabla^2 \rho_C$ (au) |
|---|---|---|---|---|---|---|
| NF$_3$H$^+$···HBeH | 20.6 | 1.114 | 0.125$^b$ | 0.051 | 0.0868 | $-0.0542$ |
| HClOH$^+$···HBeH | 18.45 | 1.157 | 0.089 | 0.042 | 0.0757 | $-0.0097$ |
| Cl$_2$OH$^+$···HBeF | 14.99 | 1.133 | 0.090 | 0.022 | 0.0789 | $-0.0191$ |
| H$_2$OH$^+$···HBeBeH | 23.13 | 1.127 | 0.111 | 0.057 | 0.0830 | $-0.0351$ |
| Cl$_2$OH$^+$···HBeH | 20.66 | 1.057 | 0.142 | 0.058 | 0.1019 | $-0.0992$ |

*Source:* Data from refs. 2 to 4.

$^a$Obtained by calculations at the MP2/6-311++G* level.

$^b$Elongation in the N–H bond.

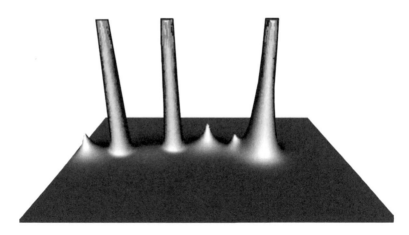

**Figure 11.1** Relief map of the electron density calculated for the charge-assisted dihydrogen complex H$_2$OH$^+$···HBeBeH shown in the plane of the HBeBeH molecule. The electron density of the HBeBeH molecule is located on the left side of the figure, and the electron density of the H–O bond, which is the proton-donating site, is shown on the right side. (Reproduced with permission from ref. 2.)

1.06 and 1.1 Å that correspond to one-half of the van der Waals sum, 2.4 Å. In addition, the lengths of O–H (or N–H) and Be–H bonds involved in dihydrogen bonding are extended considerably.

The bonding energies in Table 11.1 are in the range 15 to 23 kcal/mol after BSSE correction. These values correlate well compared to strong resonance-assisted classical hydrogen bonds, recalling assumptions regarding the partial covalent nature of dihydrogen bonding. The latter is confirmed completely by topological analysis of the electron density: The $\rho_C$ values are significant (0.076

to 0.102 au) and correspond to the concentration of the electron density within the H$\cdots$H internuclear region, and the Laplacian is strongly *negative*.

The partially covalent character of dihydrogen bonding agrees well with the calculations of energies that contribute to the total bonding energy of complexes. The most important result is that the delocalization energy, as an attractive energy term, *surpasses* the electrostatic contribution and the exchange energy. Thus, this term can be the driving force in the formation of strong dihydrogen bonds. The correlation terms were also significant. For example, the calculations performed for the dihydrogen-bonded complex $H_2OH^+\cdots H-Be-Be-H$ led to $\Delta E_{EL} = -16.90$, $\Delta E_{DEL} = -37.01$, $\Delta E_{corr} = -5.03$, and $\Delta E_{EX} = +27.61$ kcal/mol, clearly illustrating the foregoing statement. These investigations show clearly that the lengths of dihydrogen bonds can be as short as $\approx 1$ Å (versus 0.75 Å in the free dihydrogen molecule), decreasing or even abolishing the "taboo" domain shown in Figure 8.9. In other words, there is no clear border separating dihydrogen bonding and covalent bonds.

## 11.2. SPECIFIC INFLUENCE OF THE ENVIRONMENT ON DIHYDROGEN BONDING

As we have already noted, classical hydrogen bonds can be thermodynamically very strong but at the same time, easy to transform, due to fast proton transfer along a strong hydrogen bond. Such a dualism can also be seen in the dihydrogen bonding $Y-H\cdots H-X$,

$$Y-H + H-X \rightarrow Y-H\cdots H-X \rightarrow [Y-(H_2)]^+ \cdots X^- \rightarrow [Y-(H_2)]^+ + X^- \tag{11.1}$$

when hydride $Y-H$ is capable of proton transfer. It should be noted that eq. (11.1) corresponds to the simplest mechanism of proton transfer, which is probably most common. In terms of this mechanism, the H$\cdots$H unit exists as a dihydrogen bond until both of the hydrogen atoms start to move, leading to the contact ion pair $[Y-(H_2)]^+\cdots X^-$. In this context, medium-strong effects on dihydrogen bonding are particularly interesting, and some were discussed in earlier chapters. Our conclusions, here, however, are based on results obtained recently in a comprehensive study of various dihydrogen-bonded complexes of dihydride $RuH_2[P(CH_2CH_2PPh_2)_3]$ in dichloromethane, THF, THF/$CH_3OH$, and THF/$CH_3CN$ mixtures [5]. Among the dihydrogen-bonded systems investigated, the complex with $CH_3COH$ binding to the axial hydride ligand seems to be most representative.

Figure 11.2 shows a DFT/B3LYP-optimized structure typical of dihydrogen bonding with an H$\cdots$H distance of 1.721 Å, the elongated Ru–H bond donating the hydridic hydrogen (1.661 Å versus 1.629 Å for the free Ru–H bond) and the elongated O–H bond donating the proton. It follows from the data in Table 11.2 that on going from the gas phase to solution, the bonding energy decreases and the effect deepens with increasing polarity. In addition, the polarity

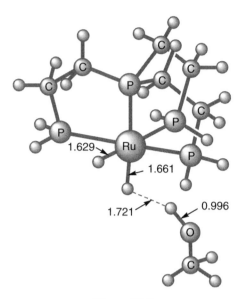

**Figure 11.2**

**TABLE 11.2. Bonding Energies, Mulliken Population Bond Values, and H···H Distances**$^a$

| Structure | Solvent | $\Delta E_{ZPE}$ (kcal/mol) | $p(H \cdots H)$ (au) | $r(H \cdots H)$ (Å) |
|---|---|---|---|---|
| 1 | Gas phase | − 8.82 | 0.023 | 1.721 |
|   | CH$_2$Cl$_2$ | − 1.49 | | |
|   | H$_2$O | − 1.01 | | |
| 2 | Gas phase | −21.91 | 0.043 | 1.592 |
|   | CH$_2$Cl$_2$ | − 5.17 | | |
|   | H$_2$O | − 3.15 | | |

*Source:* Data from ref. 5.

$^a$Obtained at the DFT/B3LYP level for the Dihydrogen-bonded complexes shown in Figures 11.2 and 11.3.

reduces the energy barrier, separating the dihydrogen bond and the contact ion pair. This effect can be assumed most common. Surprising effects have been observed for solvents that are capable of specific interactions, much as hydrogen bonding. Structural, energetic, and electronic features of a model system investigated at the DFT/B3LYP level in the presence of water are shown in Figure 11.3 and Table 11.2. It is clear that additional hydrogen bonding with an H$_2$O molecule *shortens* and *strengthens* the initial dihydrogen bond: The H···H distance decreases to 1.592 Å and the Ru–H and O–H bond lengths

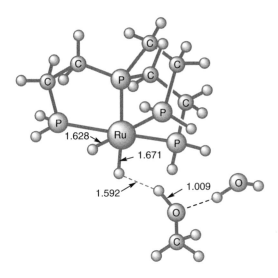

**Figure 11.3**

increase in accord with increasing Mulliken population bond value. Moreover, the bonding energy values show clearly the presence of a strong cooperative effect in dihydrogen/hydrogen-bonded systems. It should be noted that similar results have been obtained for the other systems investigated by Belkova et al. This cooperativity is particularly important in all aspects of dihydrogen bonding.

## REFERENCES

1. S. J. Grabowski, W. A. Sokalski, J. Leszczynski, *J. Phys. Chem. A* (2004), **108**, 1806.
2. S. J. Grabowski, W. A. Sokalski, J. Leszczynski, *J. Phys. Chem. A* (2005), **109**, 4331.
3. S. J. Grabowski, T. L. Robinson, J. Leszczynski, *Chem. Phys. Lett.* (2004), **386**, 44.
4. S. J. Grabowski, *Annu. Rep. Prog. Chem. Sect. C* (2006), **102**, 131.
5. N. V. Belkova, T. N. Gribanova, E. I. Gutsul, R. M. Minyaev, C. Bianchini, M. Peruzzini, F. Zanobini, E. S. Shubina, L. M. Epstein, *J. Mol. Struct.* (2007), **844–845**, 115.

# INDEX

*Dihydrogen Bonds: Principles, Experiments, and Applications*, By Vladimir I. Bakhmutov
Copyright © 2008 John Wiley & Sons, Inc.